AF704426

From:
Row By Row Bookshop
PO Box 294
Sugar Grove , NC 28679
U.S.A.

AbeBooks

To:
ABE TP ORDERS
800 AVONDALE AVE
COLUMBUS OH 43212
U.S.A.

FOR CUSTOMS ONLY

Shipping Manifest

Phone: 0000000000

Sales Order No.: 774558380
AbeBooks Purchase Order No.: 738048031
Processed: September 10, 2026 **Delivery Time:** 5 - 14 business days

Item	Author	Title	Bookseller Inventory No.
1	Dernbach, Ulrich, and William D. Tidwell	Secrets of Petrified Plants	063872
	Book Description: An ex-library copy in original pictorial hard covers of this tall-format volume. The usual ex-libris markings, and a small scuff at the top edge of the front cover. The binding is sound, and the contents are clean/unmarked. No dust jacket.		

Please retain this packing list for your records. If you have any questions about this order, please contact the bookseller using the Ask a Bookseller a Question link available in your AbeBooks account.
**Please Note: AbeBooks has collected payment for this order.

Thank you for shopping with AbeBooks.com.

AbeBooks

www.abebooks.com | www.abebooks.co.uk | www.abebooks.de
www.abebooks.fr | www.abebooks.it | www.iberlibro.com

Booksellers: Please use this page as your shipping label. Simply print the page, trim along the lines indicated above, and attach to the outside of the package making sure that the address information is visible. This shipping label can also be used for customs documentation if necessary, and we encourage you to print a second copy to enclose with the book(s).

Ulrich Dernbach
William D. Tidwell

Secrets of Petrified Plants

Fascination from Millions of Years

Authors:
Manfred Barthel, Ulrich Dernbach, Jean Galtier, Walter Jung, Hans Kerp, Robert Noll, Ronny Rößler, Gar W. Rothwell, Alfred Selmeier, Ruth A. Stockey, William D. Tidwell, Volker Wilde, W. Walton Wright

D´ORO

Impressum

Secrets of Petrified Plants
Fascination from Millions of Years

D´ORO Publishers
Germany 2002

Schlehenweg 2
D-64646 Heppenheim
Phone: +49 / 6252 / 5661
Fax: +49 / 6252 / 68587

Idea:
Ulrich Dernbach

Design and Layout:
Evgeniy Potievsky, Chemnitz

Translation:
Howard Fine, Munich

Editing Supervisor:
Dr. Ronny Rößler, Chemnitz

Image processing:
Punkt 191, Firmen- und
Produktwerbung, Chemnitz

Printing: Druck- und
Verlagsgesellschaft Marienberg mbH

ISBN 3-932181-04-2

Printed in Germany

Acknowledgements

We are particularly grateful to the authors of the scientific papers printed in this volume. Thanks to their exemplary teamwork, we were able to overcome illness and the geographic distances between individual authors so that this book could be completed on schedule.

We'd especially like to acknowledge the zeal of Prof. Walter Jung, Dr. Ronny Rößler and Prof. William D. Tidwell. This book would not have been possible without their superlative collaboration.

We benefited from and are grateful for the scientific stimuli given to the book by Dr. Hans-Jürgen Weiss, Dr. Wofgang Reichel, and Giovanni Zingo.

We cordially thank Prof. Reinhold Leinfelder from the Bavarian State Collection for Palaeontology and Historical Geology in Munich, Prof. Fritz Steininger from the Senckenberg Museum of Natural History and Research Institute in Frankfurt, and Dr. Ronny Rößler from the Museum of Natural History in Chemnitz for making available to us specimens and photos of specimens from the collections.

We are grateful to G. Burg, Tom Dominy, N. Donpon, Jim Gray, Family Dr. Haas, Ruth Jenkins, R. Jones, Ray and Dale Lyman, C. Miller, Doug Moore, J. Richmond, Wiliam Rose, Dr. Wolfgang Schwarz, D. and P. Snyder, and Ralph Thompson, and N. Webster, all of whom generously loaned us specimens for photographic purposes.

We thank Enrique Alvarez, Decio Freitas, Shirley VanDer Sluis, Ron Justman, and Chuck Scott, all of whom were helpful in our search for locations where petrified woods can be found.
We thank Fred Elsnau, Robert Noll, and Dr. Giraud V. Foster for contributing their skills and experience to create many gorgeous photos.

We thank John Gardner for making available his camera for numerous purpose.

We are grateful to Dr. Martin Antonow, Dr. Sidney Ash, Dr. Dagmar Dietrich, PD Dr. Jens Götze, Gail Matthews, Jim Mills, Gordon Morrison, Wolfgang Naumann, Dr. Petra Radehaus, Gabriele Stachow, and Jim Stichka for their proofreading of the manuscripts.

D´ORO Publishers

Preface

The authors from various parts of Europe and North America who collaborated to produce this book are all involved in scientific activities that bring them into close contact with petrified plants. In their work as professors, museum curators, or scientifically qualified amateurs, they research, collect, and interpret such fossils. They have the good fortune to work firsthand at localities of special scientific interest and to communicate the results of their research to the public in diverse ways.

Our publisher is outstandingly well acquainted with localities on every continent where silicified woods can be found. The publication of his two previous books made an important contribution towards the populari-zation of silicified plants. The materials that he collected during his extensive travels have provided top-quality specimens for research, teaching, and exhibition. For this book project, he brought together researchers who work individually at widely separated places and infected each of them with the enthusiasm that he feels for his theme.

The individual articles have been selected so that all of the major evolutionary stages of the land plants are represented: from the earliest terrestrial plants in the Devonian, through the diversity of spore-bearing plants and archaic gymnosperms in the Carboniferous and Lower Permian, into the plant kingdom of the Mesozoic, and finally to the more recent broad-leafed and needle-bearing trees of the Tertiary. Furthermore, we've concentrated on petrified plant localities that have excited general interest due to research results in the wake of new discoveries and observations made at these sites during the past few years. In most cases, the work at these areas is by no means complete, and research at some of them is still in its earliest phases. What the reader will learn here is extremely new – so new, in fact, that certain items of information in this book haven't yet appeared in the specialized scientific literature. We've also prioritized the diversity of permineralization and plant organs in our selection of fossils. Quartz variants are juxtaposed here with carbonized fossils, and entire fossiliferous chert strata are discussed alongside the predominant silicified woods.

What do these selected localities and their fossils have in common? All of these petrified plants are unusual and extraordinarily valuable fossils. Many of the specimens are large, some even as big as tree trunks, but their identification typically requires that they be scrutinized on several planes under a loupe or microscope. We're more or less helpless without optical devices, and we cannot even begin our investigations without prior preparation of the specimens. Diamond saws and polishing disks are our most useful tools in this. Lapidaries gave us a very important tool around 1830 with the invention of the thin section, which first made it possible to use the transmitted light microscope and extreme magnifi-cation on petrified specimens. The so-called "peel" technique is more elegant and requires the sacrifice of far less material. It has been used successfully for several decades on carbonate concretions, where it facilitates an extremely closely spaced sequence of cut layers.

The scientific study of petrified plants is closely associated with the history of the microscope. As early as 1665, the cellular structures of petrified plants were observed under the recently invented microscope. Our predecessors were equally grateful to use in their research all of the subsequent developments of reflected and transmitted light microscopes. Scanning electron microscopy too is currently performing a wide variety of valuable services.

All of these investigative methods, however, would be fruitless were it not for the fact that microscopically small structures of the original plants are sometimes preserved as fossils. The most important and most valuable attribute of our petrified plants is the anatomical preservation of cells, plant tissue, and entire organs as a result of permineralization, i.e. the petrifaction process. Silicium oxide, limestone, dolomite, iron sulfide and iron hydroxide, fluorite, and phosphate are the most important anorganic substances which, under special conditions, can penetrate plant tissues and preserve cellular structures. Microscopic examination of such specimens is much more complicated and much more strongly biologically oriented than corresponding scrutiny of fossilized organisms which have been preserved only as impressions or stone cores. The former field of research requires constant compari-son with the anatomy of the organs of extant plants. The study of silicified woods and coal balls is above all the study of plant anatomy!

Our primary scientific goal is to render an anatomically detailed reconstruction of entire plant organs and, whenever possible, the interrelationships between different organs. Paleozoic and Mesozoic terrestrial plant fossils repeatedly surprise researchers, who have discovered organ interrelationships which are inconceivable to botanists who specialize in contemporary plants: for example, fern fronds on woody stems, trees that grow from rhizomes, herbaceous conifers, etc. The reconstruction of complete plant communities or even entire ecosystems is another important objective. Cherts from Rhynie and coal balls from coal seams in the U.S.A. are relevant examples. Finally, under-standing the geological genesis of a fossiliferous stratum is also a scientific goal. Was the stratum originally a peat marsh, sediment at the bottom of a lake, or a deposit of volcanic ash? Where did the mineral solutions come from? How did the plants become petrified? Especially, the attempt to answer the third question repeatedly confronts us with numerous enigmas. Despite modern research methods, we still don't know the exact details of the processes which cause the permineralization of plant tissue. No contemporary researcher has ever observed the genesis of a specimen of silicified wood because geohistorical processes cannot be experimentally reconstructed. There are no simple "right" or "wrong" answers to the riddles that we strive to solve. All we can do is attempt to come progressively closer to the truth. This is one aspect which makes investigating petrified plants such an exciting undertaking – a thrilling pursuit which we invite our readers to share with us.

Manfred Barthel

Contents

Geological Timeline

marking the fossil localities
mentioned in this book
and depicting important steps in the
evolutionary history of vascular plants

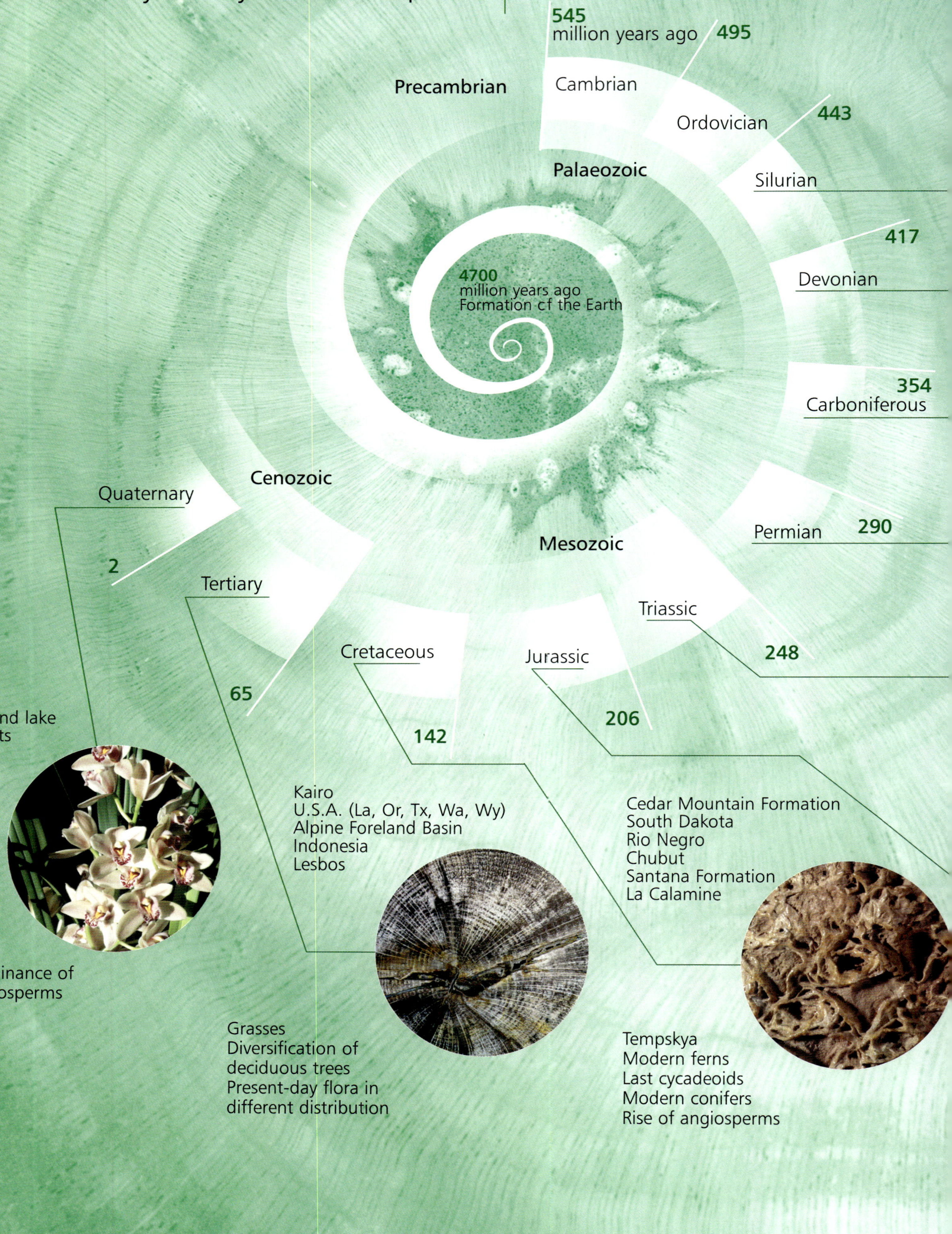

England
Barrandium
Australia

First vascular plants

Rhynie
Gilboa

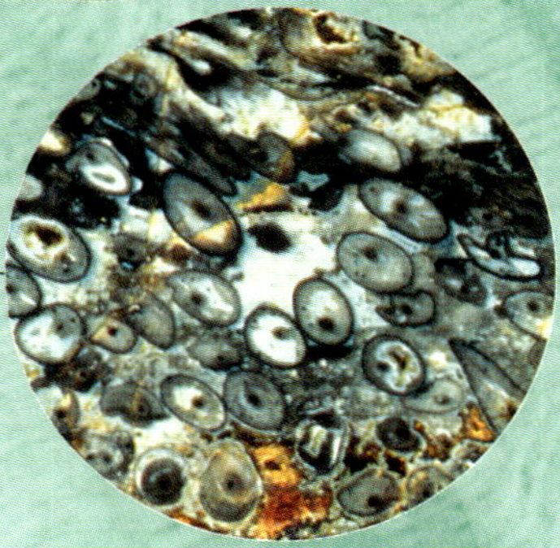

Algae
Bryophytes
Fungi
Progymnosperms
Diversification
of early vascular plants

Ohio
Coal Balls
Montagne Noire
Edinburgh
Kyffhäuser

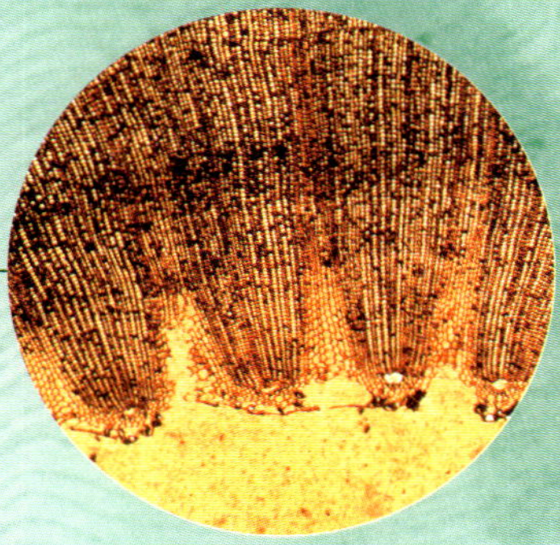

Lepidophytes
Sphenophytes
Palaeozoic ferns
Seed ferns
Cordaites
First conifers

Chemnitz
Autun
Nová Paka
Araguaina
Paraguay
Palatinate

Psaroniales
Late lepidophytes
Seed ferns
First Cycadophytes
Walchian conifers

Madagascar
Southern Germany
Zimbabwe
Brazil
Arizona

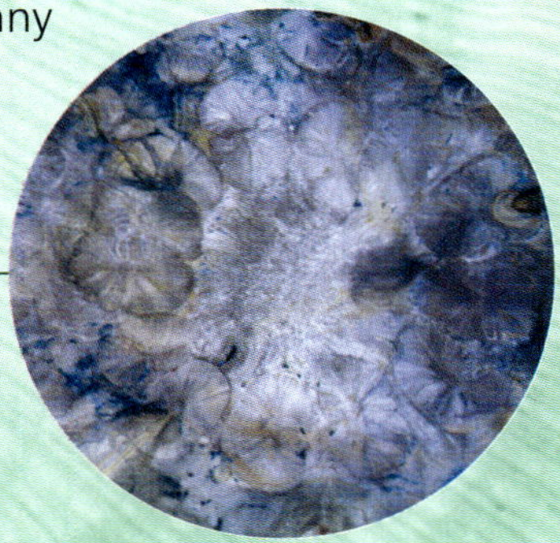

Equisetites
Filicales
Mesozoic seed ferns
Cycadophytes
Ginkgophytes
Voltziales

orrison Formation
rro Cuadrado
stralia
rkshire

Osmundacea
Nilssonia
Pentoxylon
Hermanophyton
Cycadeoids
Ginkgophytes
Conifers dominant

The circles show details
of the following plants:

Quaternary:	*Cymbidium* sp.
Tertiary:	*Quercoxylon* sp.
Cretaceous:	*Monanthesia magnifica*
Jurassic:	*Osmundacaulis* sp.
Triassic:	*Rhexoxylon africanum*
Permian:	*Medullosa stellata*
Carboniferous:	*Arthropitys communis*
Devonian:	Rhynie Chert
Silurian:	*Cooksonia pertonii*

RONNY RÖßLER

Methoden für die Untersuchung strukturzeigender Pflanzenorgane

Study Methods for Determining the Structure of Plant Organs

In the majority of cases, collectors and researchers of petrified woods occupy themselves with two central questions:

(1) First, one wants to understand the botanical behaviour and taxonomic position of the wood. Microscopy is the most important diagnostic method here.

(2) Secondly, questions arise anew concerning the permineralization process that ensued at each individual deposit. Palaeobotanists, i.e. those who study fossil plants, rely on a variety of modern analytical methods of physics and chemistry in their efforts to answer these questions.

From a Find in the Field to a Piece of Scientific Evidence

Considerable effort is required in order to answer the first-mentioned question. Only in a few cases does the macroscopic appearance of a permineralized wood specimen suffice to support its immediate identification. First, the majority of specimens must be subjected to rather lengthy and elaborate preparatory work before their identity can be determined. Furthermore, a precise knowledge of the geological situation and the age of the strata in which the specimen was found can be very helpful. This knowledge can help an investigator decide by whom he or she would be best advised. Just like physicians, engineers, or lawyers, paleobotanists tend to specialize. There's no palaeobotanist who can cover the complete field of palaeobotany. Most researchers specialize in particular taxonomic groups or temporal segments of geological history. The older the fossil, the less helpful comparison with recent plants will be. Hence, the most accurate possible documentation of the discovery site and the conditions surrounding the find is always the correct first step. Failure to record relevant data at the outset makes it considerably more difficult to later identify an isolated specimen that may seem to have appeared out of nowhere.

To begin with, each researcher must answer the question of how far he or

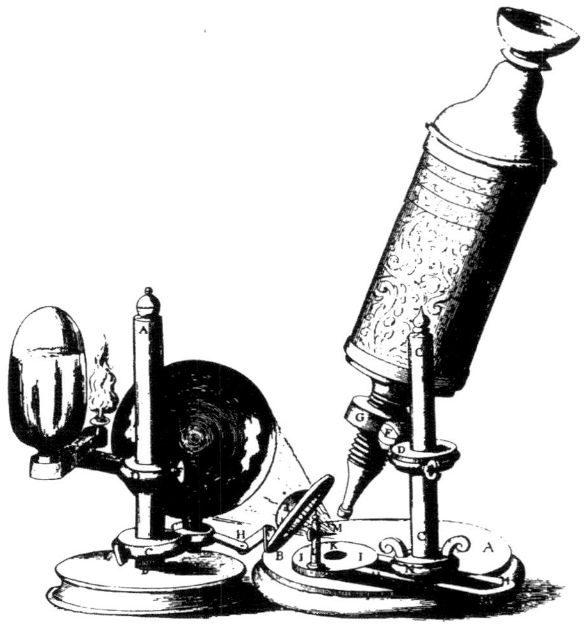

1 Microscope of Robert Hooke
shown in his book "Micrographia" in 1665

she wants to proceed with the identification. Whereas an identification down to the level of the genus usually remains a more or less straight-forward project, further classification down to the species level is almost always a task for a specialist. But to the surprise and disappointment of many collectors, even a specialist can scarcely be expected to complete the identification, as it were, on his feet and while standing on the threshold of the collector's door. It often takes a considerable amount of time to reliably identify a specimen of fossilized wood. In accomplishing this task, we can rely less upon the scanty classificatory literature than upon our many years of experience in collecting, observing, and documenting our own finds, and, as our most important tool, comparison with similar identified specimens. Ongoing scientific exchange among fellow collectors, friends, or paleobotanists, perhaps within the framework of lectures or workshops, is an important aid. With the exception of a few innovations and refinements, the tools, equipment, and methods used to identify permineralized woods have remained rather classical. One begins by examining the breakage planes or relatively simply prepared polished surfaces from a cross-section of the woods with the help of a magnifying lens. Although the limits of this method are very quickly reached, this procedure has remained the first step on the path towards identification. **Microscopy** is a more powerful method and one which is inseparably associated with the investigation of fossilized woods. A silicified wood was one of the first objects that Robert Hooke subjected to examination in 1665 under the lens of his newly invented microscope. Hooke, thus, became the first person to furnish the important proof that fossilized wood is indeed a relict of earlier life. He illustrated his observations and compared them with sketches of recent wood in his book Micrographia.

This was the birth of **reflected light microscopy**, which has remained the method of choice as long as the contrasts within the tissue are sufficiently strong. Typically examined objects include the polished surfaces of the woods and/or their fragments. A whole series of important classificatory criteria can be used here.

Transmitted light microscopy, an alternative method, has been in use since 1830. The basis for this method was provided by Henry T. M. Witham of Lartington. Basing his research on methodical preparatory work carried out by the optician William Nicol, Witham observed permineralized woods by using **thin section samples** under a transmitted light microscope and thereby gave palaeobotanists access to an entirely new and impressive world. Numerous scientific works which exploited these revolutionary methodical foundations were subsequently published in England and Germany during the 1830's and since.

How does one make a thin section? The first step is to grind a perfectly even surface on the area of the speci-

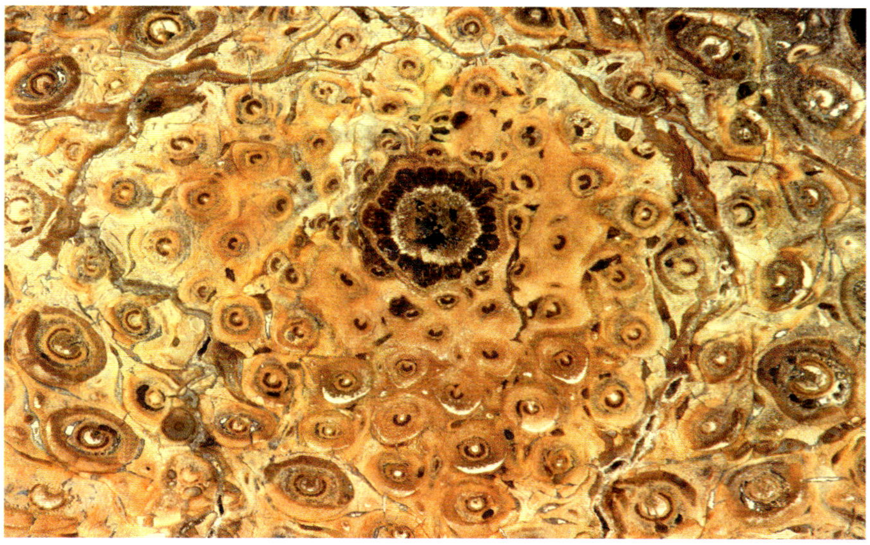

2 Transverse section
of an osmundaceous fern
Tertiary, Trimmelkam, Austria
Reflected light image of a polished surface
Museum für Naturkunde Chemnitz, 3 x

men of silicified wood one wants to examine. The simplest way to accomplish this is by hand, although now the grinding is usually accomplished by machines using a rotating grinding wheel with either loose or fixed abrasive grains of progressively smaller sizes, culminating in the finest grains for the smoothest surface. Grains of silicon carbide provide a relatively favorable cost-labor ratio. The next step is to affix a glass slide to this ground surface. Adhesives commonly used for this purpose include Canada balsam, thermoplastic quartz cement, and epoxy resin. The first two of these embedding media have the advantage that the polished specimens can be re-embedded if and when this should ever become necessary. Epoxy resin, which is primarily used in professional laboratories, has two disadvantages: the embedding is permanent; and it can seldom be achieved entirely free from the bubbles which are often inadvertently created when the two components of the adhesive are mixed. Each researcher empirically develops his or her own tricks and knacks, e.g. previous warming of the materials, the use of vacuum chambers, or other practical methods. After the glass slide has been glued to the polished surface, as thin a slice as possible is cut from the specimen along a plane that parallels the slide. The freshly cut surface of the specimen is then ground 5 or 6 times using progres-sively smaller-grained abrasives until the desired final thickness is reached. This thickness, or rather thinness, varies from slide to slide because it depends upon the specific attributes of the material and its degree of transparency. Typical thicknesses for palaeobotanical thin sections range between 30 and 100 microns, which makes them significantly thicker than the thin slides or thin sections which are typically used by mineralogists. Frequent checks during the polishing process have shown themselves to be worth-while. For our purposes, the upper surface of the thin section is left without a covering slide. On the one hand, the absence of such a cover allows the thickness to be reduced still further, if this should be needed later; on the other hand, the exposed specimen

3 Thin section preparation from the collection of Johann Traugott Sterzel
Museum für Naturkunde Chemnitz

remains available for follow-up investigations, e.g. via cathodo-luminescence microscopy (CL) or scanning electron microscopy (SEM). Thin sections have been numbered among the standard tools of palaeo-xylotomy since the mid 19th century, despite the fact that their manufacture requires a fair amount of time, experience, and manual dexterity.

In many cases, a single thin section is not sufficient to support a positive identification because silicified woods frequently display a certain degree of variability in their diagnostic characters. The best example of this is provided by the palaeophytic conifers and/or cordaites woods, which are notoriously difficult to differentiate from one another.
A great deal of credit in this field is due to H.R. Göppert. At a very early date, his studies included comparisons between the anatomy of these ancient woods and that of the wood found in recent conifers. He was also the one who first emphasized that in order to identify a specimen of silicified wood, "...three different polishes and three different cuts are necessary": one transverse section and two longitudinal sections, of which one should be parallel to the outer surface and the other parallel to the rays. This has remained the usual research practice to the present day. Göppert had his "Arboretum fossile" (Fig. 7), a series of 70 thin sections from various silicified gymnosperm woods, produced in series for the purposes of study and comparison.
In modern studies there is applied a material-saving optimization of the thin-section technique. The specimens are cut into slabs, and so called **"peels"**, cellulose acetate peel preparations, that produce an ultrathin section of the plant material, are made from each ground and etched cut surface. This method, however, is unfortunately limited to materials which contain sufficient amounts of organic remains, e.g. the cell walls of

4 - 6 Transmitted light images of Cordaites wood
Permian of Palatinate (transverse, radial, tangential)
1 x 1.5 mm, photo: Noll

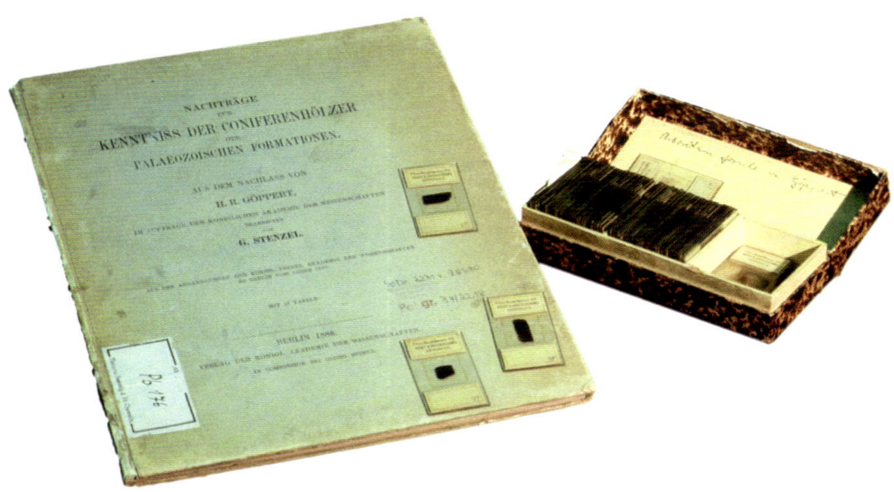

7 "Arboretum fossile", collection of thin section preparations made by H.R. Göppert around 1880, and given to scientific institutions
Museum für Naturkunde Chemnitz, photo: Noll

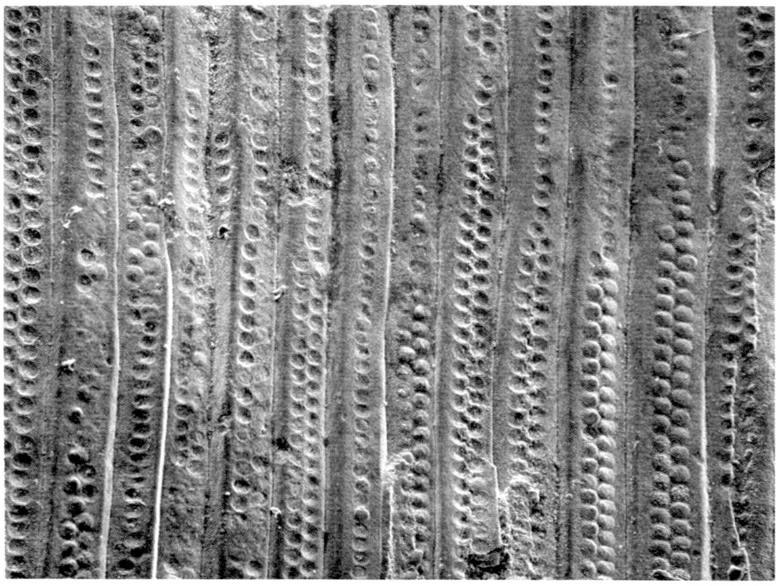

8 SEM image of a Permian conifer wood showing pits of the radial tracheid walls
width 0.5 mm, photo: Wilde

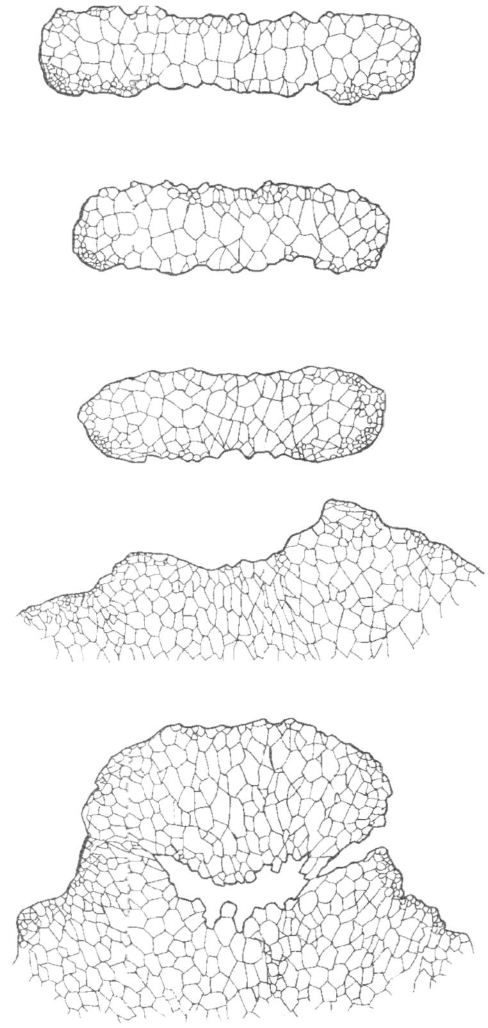

9 Drawing with the aid of a camera lucida showing the ontogenetic change of the leaf trace xylem configuration of the tree fern *Grammatopteris freitasii*

the plant tissue. Specimens of silicified wood which have been almost completely permineralized do not yield useable peels. The strength of the peel method was demonstrated above all through the contributions it has made to the investigation of Upper Carboniferous coal-ball floras (Rothwell, in this volume). The carbonate concretions often contain a large number of non-compacted plant remnants which can be interpreted as in situ peats and which even support quantitative research. The etching can be accomplished relatively safely with a diluted solution of hydrochloric acid. After the foil is peeled off, the specimen is polished again. In this way, a series of peels can be produced without sacrificing a significant amount of the original specimen. With some practice, these prints can be used to facilitate the reconstruction of the plant's internal anatomy, the relationships between its various organs, and its branching patterns. The invention of the peel method gives scientists access to diversely useable and comparatively sturdy preparations which have proven to be quite suitable for comparative purposes. Details in the peel can also be cut apart, embedded, and studied using the same techniques as are ordinarily used to prepare microscopic slides.

Modern investigation methods, like **scanning electron microscopy** (SEM) invented during the 20th century enabled studies at higher magnifications. These methods, however, are available only to specialists because the equipment required to perform them is prohibitively costly and elaborate. The specimen is first affixed to a metallic stub. The necessary electrical conductivity of the specimen's surface is achieved by the application of vaporized carbon, gold, or some other element. In the majority of

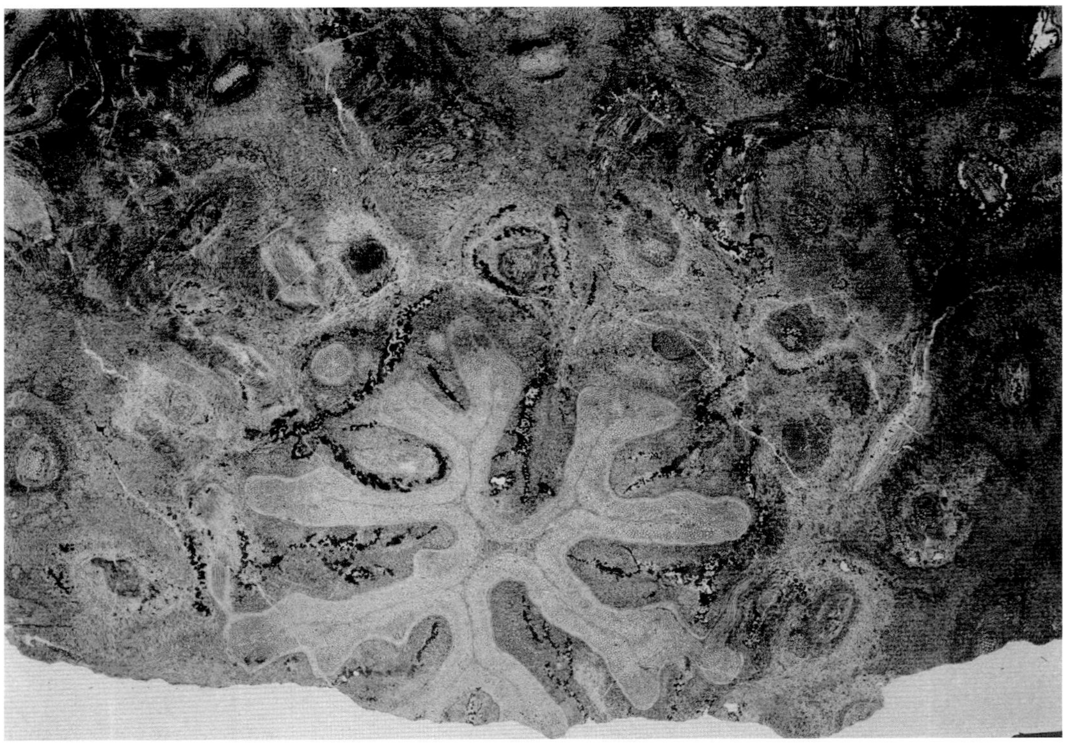

10 Transverse section of the small tree fern *Asterochlaena laxa* from the Permian of Chemnitz

Note the spirally arranged leaf traces around the central actinostele
4 x
Historical photo made by Paul Bertrand in 1911

cases, tiny fragments are sufficient to provide the desired information about the tissue and cells. If miniscule polished pieces are used, then the surface is additionally etched. Hydrofluoric acid and/or their vapors are used for siliceous materials, a method which requires specially ventilated fume hoods and the greatest caution. Unlike light-optical methods, electron microscopy, that is based on preparations of broken or etched surfaces, yields further information about internal anatomy of permineralized woods.

Scientific drawings and **photography** are indispensable aids to interpretation for all microscopic methods. Not until an observer begins to draw a specimen is he or she obliged to thoroughly consider each sketched line and thus the studied object itself. Furthermore, scientific drawing is not an art, but a learnable craft. In any case, it's a necessary step towards the overcoming of "picture book palaeontology." Most binocular microscopes support equipment with so-called drawing mirrors (camera lucida) which render eminently useful service. In exceptional cases, one can also draw a sketch based upon a photograph, although this method cannot exploit the important advantages which can be derived from adjusting a binocular microscope's depth of focus. As far as photography is concerned, we can learn a great deal from our predecessors. Today too, I'm repeatedly astonished to see the high quality and fidelity to detail that are so unmistakably evident in their images (Fig. 10). Methods which support direct illumination remain unexcelled with respect to the sharpness of the images produced. Macro lenses attached in front of the camera are very serviceable up to a scale of nearly 1 to 1. Unfortunately, our binocular microscopes are optical compromises in which we necessarily sacrifice sharpness, despite all adjustments of aperture. Qualitative losses are frequently unavoidable with lower magnification microscopes at magnifications from 2:1 to approximately 10:1. The sharpness of the image again becomes far more acceptable at magnifications between 10:1 and 100:1 under the stereo microscope.

It is here that the lion's share of the work of identifying fossilized woods is done. Several years ago, digital photography began to make inroads into paleontology. Not only does this innovative technology offer excellent opportunities for shooting photos as far as the use of available light is concerned, it also provides a wide spectrum of options for the subsequent processing of the images.

How Can we Solve the Riddle of the Genesis of the Fossilized Woods ?
There is no such thing as "the" path towards petrification. Not only is there "more than one way to skin a cat," there also seems to be more than one way to petrify wood.
A whole series of minerals have been observed in fossilized woods, but the process itself remains fraught with riddles. If one wants to know more about the process which led to the fossilization of a particular specimen, comprehensive solid state analyses are indispensable. Palaeontologists are therefore frequently obliged to rely on interdisciplinary collaboration with physicists, materials experts, and chemists. Chemical analyses of petrified specimens have long been important, but they require the rather complicated decomposition of the tested material. Furthermore, their results, which provide information about the average composition of the specimen, are unsatisfying because they scarcely pay any attention to the role played by trace elements in causing the coloration of our silicified woods. For these reasons, the more common practice for determining the mineral composition of "petrified" woods (chief components, auxiliary components, and accessories) continues to rely on modern analytic methods which can provide scientists with information about the distribution of individual elements and compounds. In recent years, the methods of choice have been the **electron beam micro-probes** with wavelength dispersive spectrometers or scanning electron microscopes with energy dispersive X-ray analyzers. These instruments collect characteristic X-ray radiation from the specimens. The characteristic wavelengths and energies of the radiation reveal the iden-

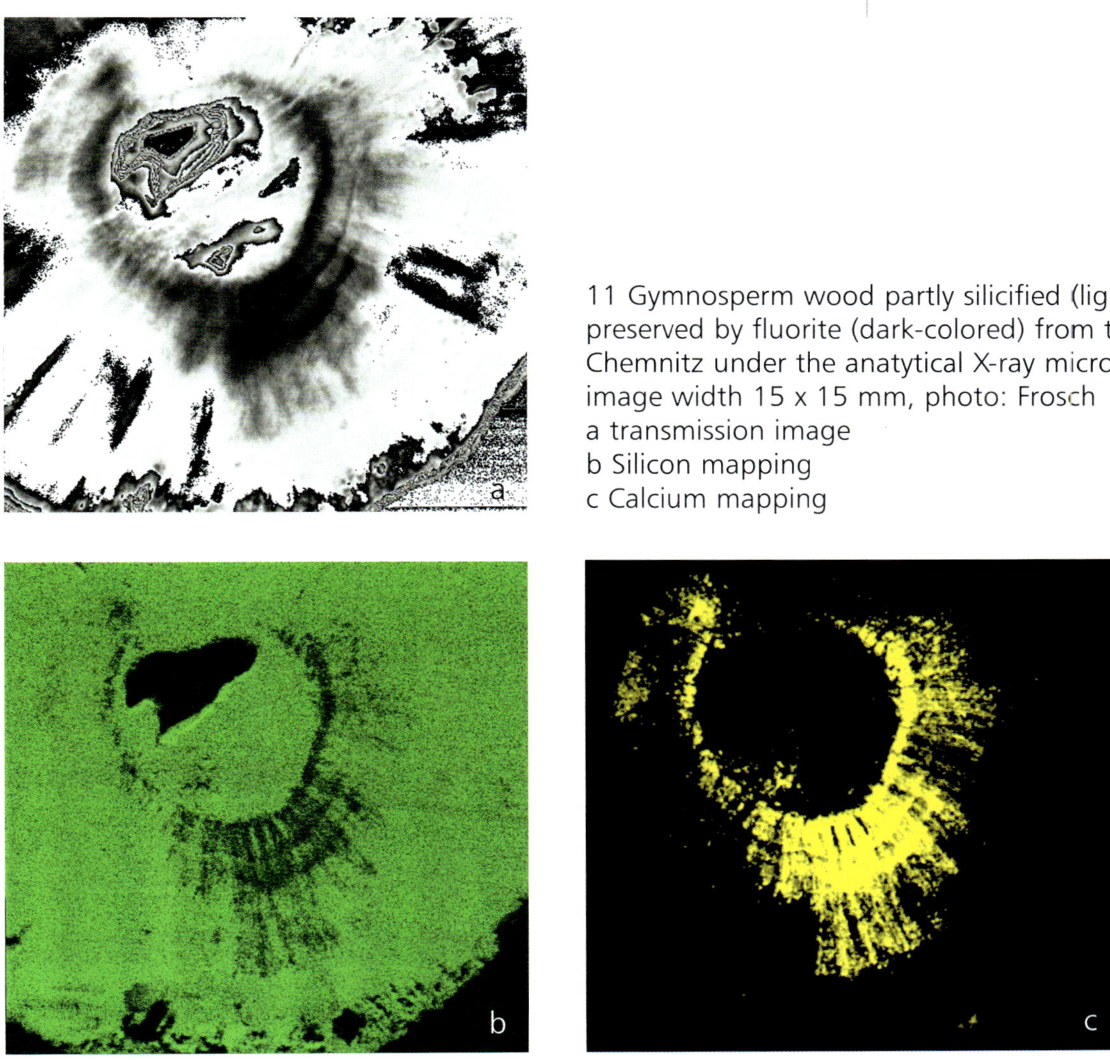

11 Gymnosperm wood partly silicified (light-colored) or preserved by fluorite (dark-colored) from the Permian of Chemnitz under the anatytical X-ray microscope
image width 15 x 15 mm, photo: Frosch
a transmission image
b Silicon mapping
c Calcium mapping

tity of the atoms within the samples. By scanning with the fine electron probe and mapping the characteristic X-rays, we can generate element-distribution images for elements ranging from boron to uranium. Small samples are necessary for this technique, each of which is generally only a few centimeters in size, and must be first prepared by cutting and polishing before their surfaces are coated with a thin conductive layer.
The excitation of the characteristic X-ray radiation is possible not only by means of electrons, but also by means of high-energy X-ray irradiation in that case called fluorescence. Most recently, experimental procedures have been devised which make it possible to focus the X-ray beam to a diameter of just a very few micrometers. If a specimen exposed to a beam of this sort is shifted on an x-y table, element-distribution images can be generated. A device of this kind called an **analytic X-ray microscope**. Three advantageous features of these devices are: there are virtually no limitations on the size of the samples; the previous application of an electrically conductive coating is unnecessary; and the technique does not need a vacuum and can be employed under ordinary atmospheric conditions. Laborious preparation of the samples is no longer necessary and the sample itself survives the experiment unscathed. One limitation of this technique is that it can only analyze at atomic number 11 (sodium) or higher on the period table of the elements.

MicroRaman spectroscopy is a very modern method of solid state spectrometry in which excitation is produced by a laser beam having a wavelength in the range of visible light. The high intensity of the laser beam and its narrow diameter of just a few micrometers, in combination with state-of-the-art devices and modern computer technology, support both the qualitative identification of the minerals present, as well as their distribution. Raman spectrometry measures the energy shift of the radiation, which depends upon the specific geometric arrangement of the atoms with respect to one another; and yields a spectrum with characteristic bands whose energy positions are given as reciprocal wavelengths. This makes it possible to differentiate between the various spatial arrangements of SiO_4 tetrahedrons in the polymorph forms of SiO_2 such as a pha-quartz, cristobalite, tridymite, or chalcedony. For carbon, which is usually the last remnant of organic substance in a specimen of silicified wood, it's possible to gather information about the dimensions of individual areas of the crystals or to identify specific forms such as anthracite or graphite, i.e. to recognize various states of coalification.

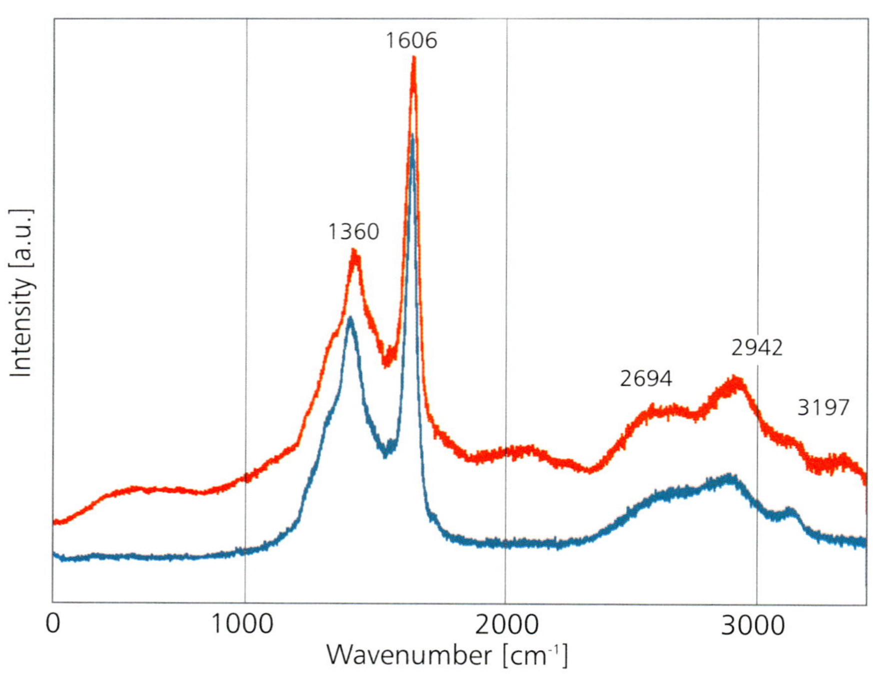

12 RAMAN-spectrum of the remaining anthracitic particles in the sclerenchyma sheath of a *Psaronius* root (red) in comparison to an anthracite calibration sample (blue)

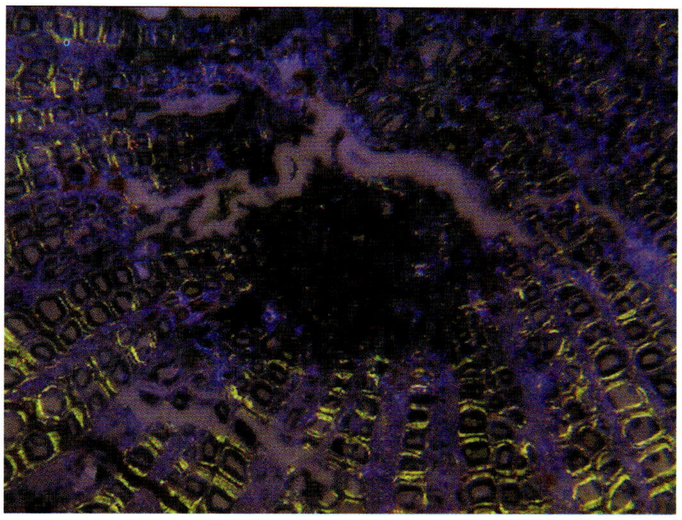

13 Cathodoluminescence image of a star-shaped vascular bundle of *Medullosa stellata* from the Permian of Chemnitz
The yellow parts reflect the first mineralization phase; the blue parts show favoured infiltration ways (rays) of the permineralization agent (blue) during a second mineralization phase.
width 1.6 mm, photo: Götze

Cathodoluminescence has shown itself to be a useful means of elucidating the permineralization process per se. Here, the investigator attempts to establish the presence of various SiO_2 phases or generations, for example, the existence of multiphase processes during the course of the permineralization. The cathodoluminescence method requires polished and vaporcoated thin sections as its objects of study. Physicists explain luminescence phenomena as follows: The addition of energy causes electrons to be lifted out of their valence bands and into higher energy levels; afterwards, when they fall back into a lower-energy level, they release their characteristic luminescence. The combination of an electron gun with a conventional light microscope makes it possible to spectrally resolve and directly observe these luminescence phenomena in microscopic dimensions. One particularly successful experimental procedure is to allow a beam of electrons to excite the conductive side of the sample in a vacuum, thus making it possible to observe the cathodoluminescence through the thin section on the front side of the sample. The tremendous analytic potential of luminescence analysis lies in the fact that the specific luminescent capacity of particular minerals reflects the presence of disturbed regions in the crystal lattice, i.e. it can render visible the actual structure of the mineral. This, in turn, makes it possible to arrive at conclusions about particular structural attributes and conditions of genesis for the individual mineral phases.

One Example Among Many

The complex application of the aforementioned analytic methods has recently made it possible to learn the following facts about fossilized woods from the Permian of Chemnitz:
The SiO_2 matrix of the silicified woods primarily consists of alpha-quartz, with subordinate constituents of cryptocrystalline chalcedony.
Organic remains from the original plant tissue are present as coalified particles. Structurally, we're dealing with anthracite here.
The accessories which cause the coloration in the SiO_2 matrix consist primarily of ferric, especially Fe_2O_3, and ferrous compounds with subordinate constituents including cobalt, copper, and nickel.
The silicification very probably occurred as a multiphase process.
There do not seem to be any significant differences between the various plant genera and types of tissue.
It was possible to reconstruct the permineralization process in the following partial steps:

1. Primary silicification and/or partial fluoritization
2. Subsequent coverage resulting from secondary hydrothermal processes and/or secondary SiO_2 mineralization, when the fluorite was partly displaced by quartz
3. Precipitation of secondary cements.

HANS KERP

Der Rhynie Chert - das älteste und am vollständigsten erhaltene terrestrische Ökosystem

The Rhynie Chert - The Oldest and Most Completely Preserved Terrestrial Ecosystem

The earliest land plants are circa 420 million years old. These were leafless axes, only few millimeters to a few centimeters tall, which are preserved as compressions and impressions, occasionally still with coaly material still attached. Most Late Silurian and Early Devonian floras show this preservation.

The flora of Rhynie, a small village in Aberdeenshire, Scotland, is a unique exception. In 1912 the Scottish geologist William Mackie discovered an occurrence of Early Devonian plant-bearing cherts, which have been known ever since as the "Rhynie Chert". The Rhynie Chert contains the oldest anatomically preserved land plants. Moreover, these cherts have also yielded a broad spectrum of animal remains, including a few crustaceans, several trigonotarbids (spider-like arthropods), mites, and a springtail species, which together constitute the oldest known well-preserved terrestrial fauna.
The Rhynie Chert, which is dated at 396 Ma, is now considered to be one of the world's most important fossil deposits. The local outcrop conditions are extremely poor. Initially only some loose blocks of chert were found and much later a few temporary trenches were made. The first comprehensive studies on Rhynie Chert plants were published by Kidston & Lang from 1917 to 1921. Because of its special importance, the Rhynie locality has been declared a Site of Special

1 A view on Rhynie from the South with the chert occurrence West of the village

Scientific Interest (SSSI) and is only accessible for scientific research. Although the flora and fauna have been investigated for over eighty years, Rhynie Chert research still yield new and often unexpected results.

The chert-bearing deposits are intercalated between volcanic rocks and have been formed as sinters by the deposition of silica in hot springs. Several types of chert can be distinguished. Thick-bedded cherts with complete plants preserved in life position were formed as a result of occasional flooding by silica-rich water. Silicification processes must have occurred very rapidly because often even the finest details (e.g. germinating spores) and short-lived stages (e.g. various developmental stages of particular fungi) are perfectly preserved.

Not only the anatomical details of the higher land plants in this flora can be studied, but in several cases also their entire life cycles have become known. Apart from vascular plants, many types of fungi, algae, and even the oldest lichens have been described from the Rhynie Chert; interactions between fungi, land plants, and animals can be demonstrated.
The very diverse group of fungi includes saprophytic forms which take part in the decay of organic matter, fungi which live in symbiosis with plants, as well as parasitic forms. Although their producers are often still unknown, a wide variety of coprolites provides evidence of a broad spectrum of animals ranging from pure vegetarians to real omnivores. Along with fungi, animals too played an important role in the decay of organic material and in soil formation.

Thus far, six vascular land plants have been described from the Rhynie Chert. These plants were relatively small, only 10 to 40 cm high. Their aerial axes were usually naked and dichotomously branched with terminal or laterally attached sporangia having thick, multi-layered walls. Each of the Rhynie Chert plants had developed its own growth form, indicating that the individual species were well adapted to specific habitats.

Rhynia gwynne-vaughanii and *Aglaophyton major*, which are traditionally assigned to the Rhyniophytes, look very similar but differ in size and in their water-conducting xylem cells. *Aglaophyton* is undoubtedly the best-known plant from the Rhynie Chert. This species was up to 20 cm high and had leafless, creeping axes which bifurcated repeatedly and bent upwards to form aerial axes. Particularly the basal parts of the aerial axes had numerous lateral axes which could be interpreted as vegetative daughter plants. The entire plant was apparently lying on the substrate, which is evidenced by the presence of stomata on the lower surfaces of the in situ preserved lying and creeping axes. The simple stomata consist of two kidney-shaped guard cells. The *Aglaophyton* axes borne terminal fusiform sporangia. With their creeping axes, single individuals of *Aglaophyton* could occupy comparatively large areas.

2 A piece of Rhynie Chert with several horizons with *Aglaophyton major*
height 13 cm

3 A thin section with *Aglaophyton major*
width 2.3 cm

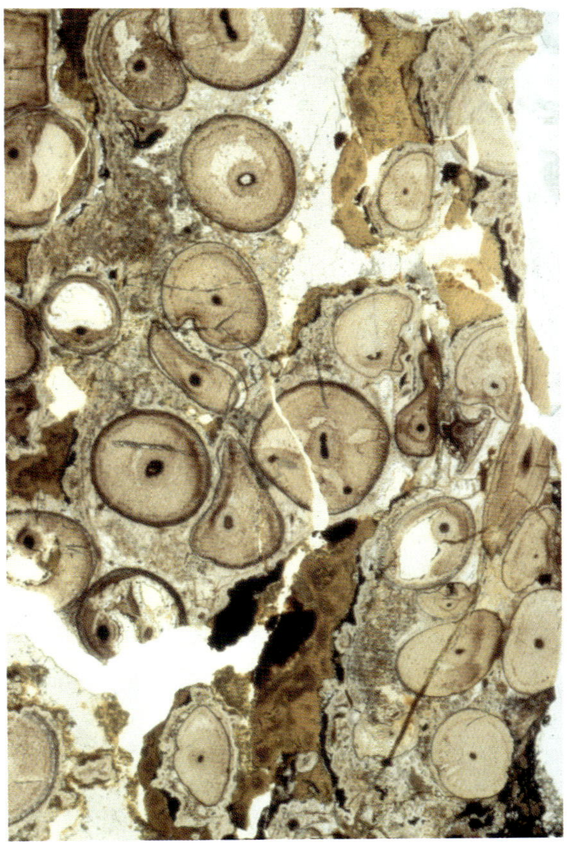

However, *Rhynia gwynne-vaughanii* had a rather bushy growth form. Two other species of Rhynie Chert flora are also quite well known. These two taxa have subterranean rhizomes and leafless aerial axes. *Horneophyton lignieri* has bulbous rhizomes, while the underground parts of *Nothia* aphylla consist of numerous, successive generations of axes. These subterranean parts of the plant grew horizontally, constituting a system of axes that formed a large rhizome. The upper parts of the axes grew upright and consisted of repeatedly bifurcated aerial axes.

Asteroxylon mackiei is the only Rhynie Chert plant that had leaf-like structures which, however, cannot strictly be called true leaves due to their lack of a midvein. This species is one of the earliest lycopsids. This plant was up to 40 cm high and its creeping rhizomes had repeatedly bifurcated roots which delved as much as 15 cm deep. The leafy aerial axes were bifurcated but also had additional lateral axes. The "leaves" are up to 5 mm long and more or less spirally arranged. The center of the axis consists of a characteristic, massive, cruciform to star-shaped xylem which sends out lateral bundles

4 *Aglaophyton major*: vascular strand, width 800 μm

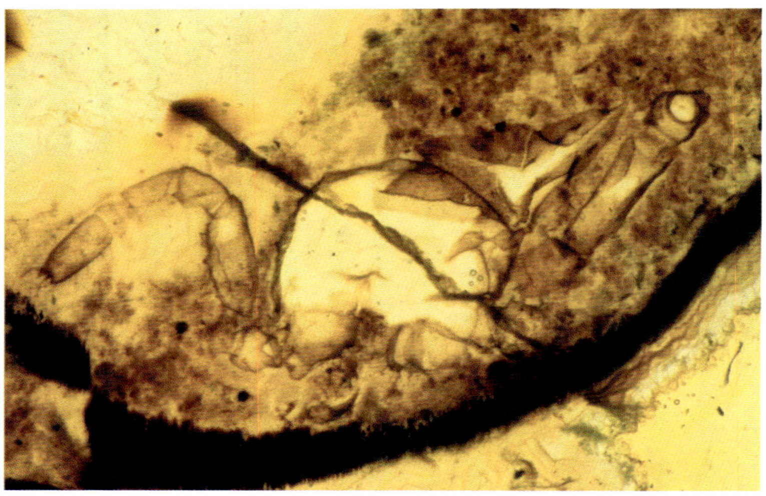

5 A trigonotarbid (spider-like animal) inside a sporangium of *Aglaophyton major*, width 5 mm

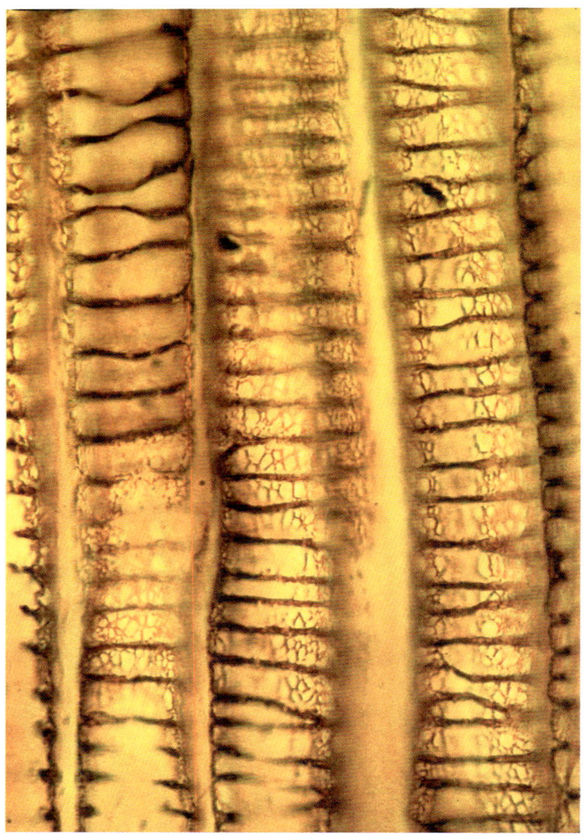

6 *Asteroxylon mackiei*: Tracheids (longitudinal section), width 130 μm

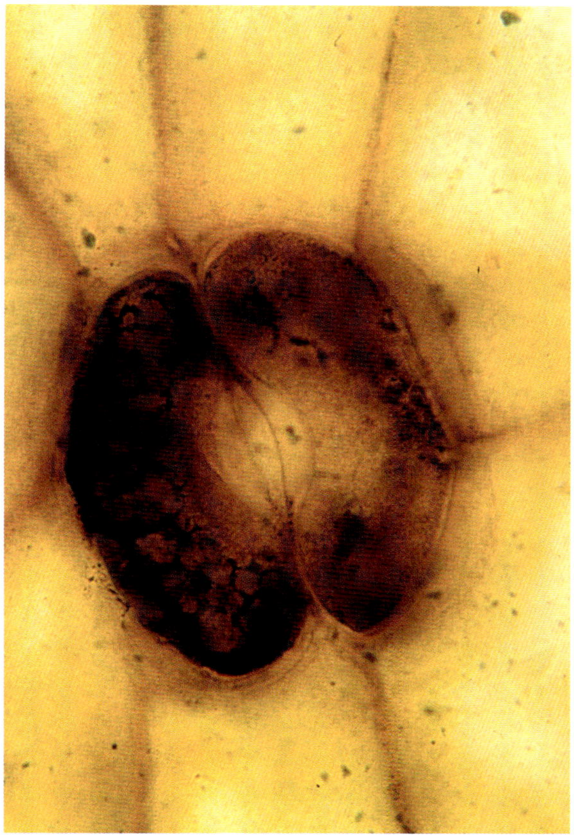

7 *Aglaophyton major*: a stoma in surface view, width 110 μm

8 *Horneophyton lignieri*: three spores of a tetrad, width 65 μm

leading toward the "leaves", but ending at the axis margin without entering the "leaves". The sporangia consist of two valves standing on a short axis, scattered between the "leaves". In comparison to other Rhynie plants, the organization of *Asteroxylon* is rather complex. With its deeply penetrating roots, it could also absorb water from deeper soil layers, thus making this species less dependent than the other species upon the moisture content of the soil surface. The rather complex stele (conducting system) guaranteed a much better stability that could be achieved with minimal effort. In this context, one can compare the stability of a one-meter-long massive iron cylinder with that of a T-beam of the same length and weight.

Although the number of land plants known from the Rhynie Chert is relatively small and the systematic position of some species has not yet been satisfactorily solved, the Rhynie Chert flora already ranks among the best-known fossil floras. Even if the organization of the early land plants seems at first glance to be not very complex and several taxa may even appear very similar, detailed studies demonstrate that ecological differentiations in morphology and anatomy resulted in a variety of different life strategies soon after the colonization of the land.

The earliest land floras are of the greatest interest from an evolutionary point of view. The Rhynie Chert, provides which is one of the very few almost completely preserved early terrestrial ecosystems, provides a unique picture of the early phase of life on land.

JEAN GALTIER

Außerordentlich gut erhaltene Pflanzen aus dem Unterkarbon

Exceptionally Preserved Lower Carboniferous Plants

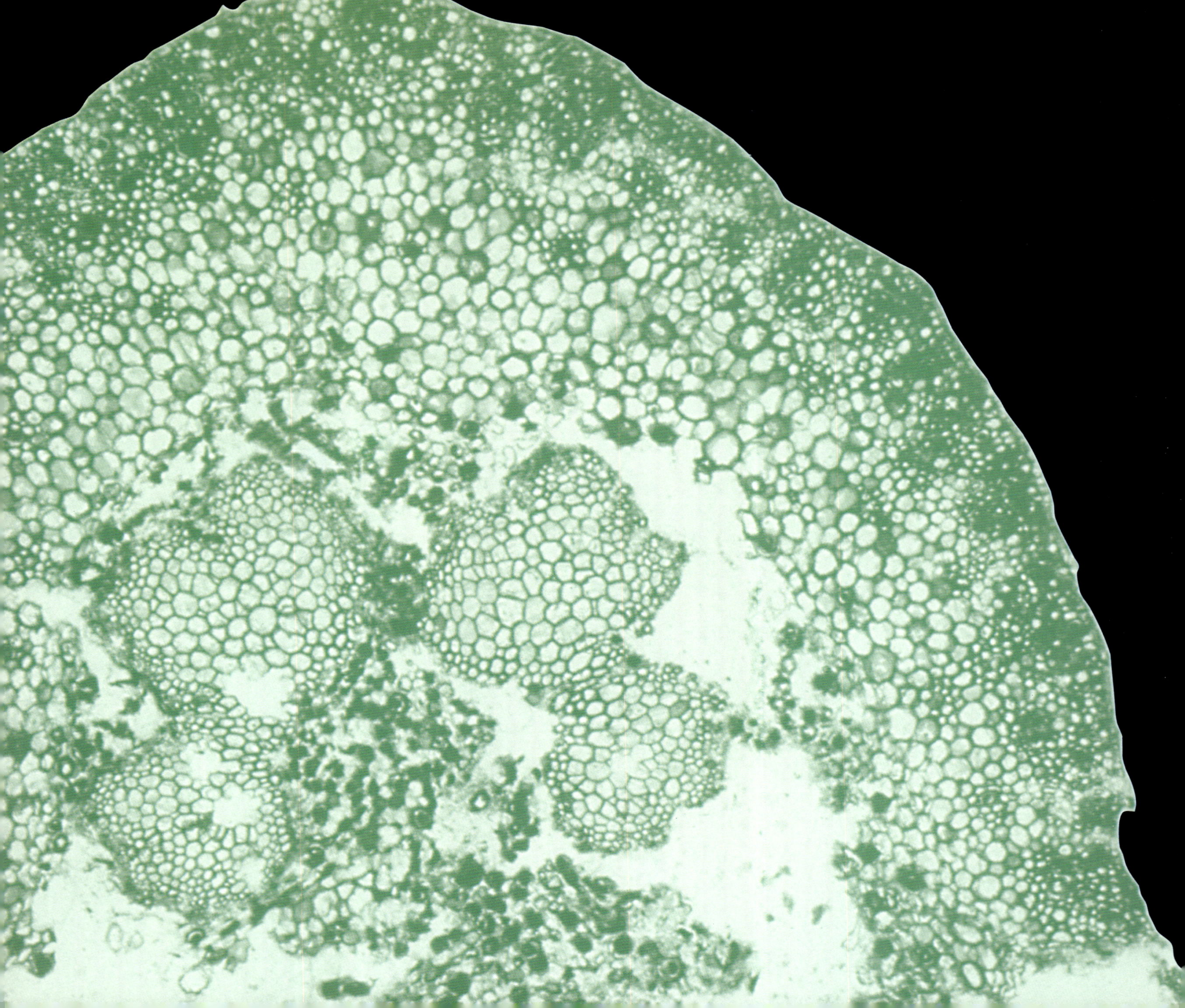

The Lower Carboniferous (beginning about 360 millions years ago) was a time of rapid evolution of plant form when, for the first time in the history of the plant kingdom, five major groups of plants attained the tree habit and when plants reproducing by seeds became very diverse.
The best available data on the organization, development and ecology of these organism is revealed by anatomically preserved plants. These have been intensively studied in Western Europe, particularly in Great Britain, Germany and France since the 19th century.
Several tens localities have yielded diverse fossil floras which are found as in situ communities or as drifted assemblages in three main settings: volcanic deposits, fluvio-lacustrine sediments and marine sediments. It is an opportunity to present examples of fossil plants of Lower Carboniferous age found in volcanic environments.
Plant deposits of Late Viséan age (about 330 million years ago) from Esnost, near Autun in the northern Massif Central of France, were discovered by Renault in 1893-96. In this locality the plants are found in angular cherts associated with anthracitic coal beds. They occur as lenses within a series of rhyolitic lavas which bound the northern margin of the Upper Carboniferous-Permian basin of Autun. The rhyolitic volcanics represent a widespread facies (also called "tuf anthracifère") in the northern part of the Massif Central. At Esnost the internal structure of the plants is preserved by silica and represents an excellent example of the permineralisation process which involves the infiltration of cellular and intercellular spaces with minerals shortly after burial.

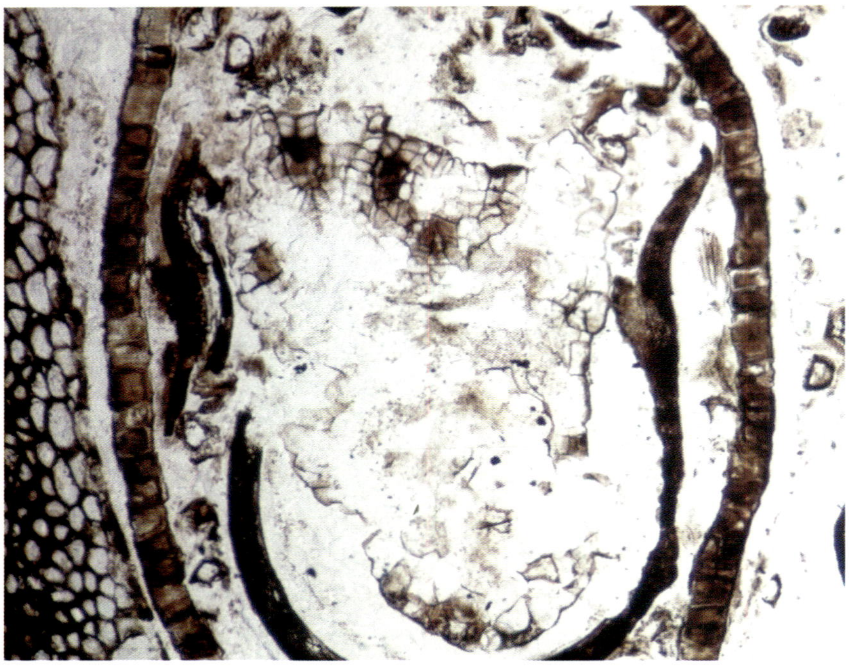

1 *Lepidodendron esnostense*
megaspore with female gametophyte
100 x

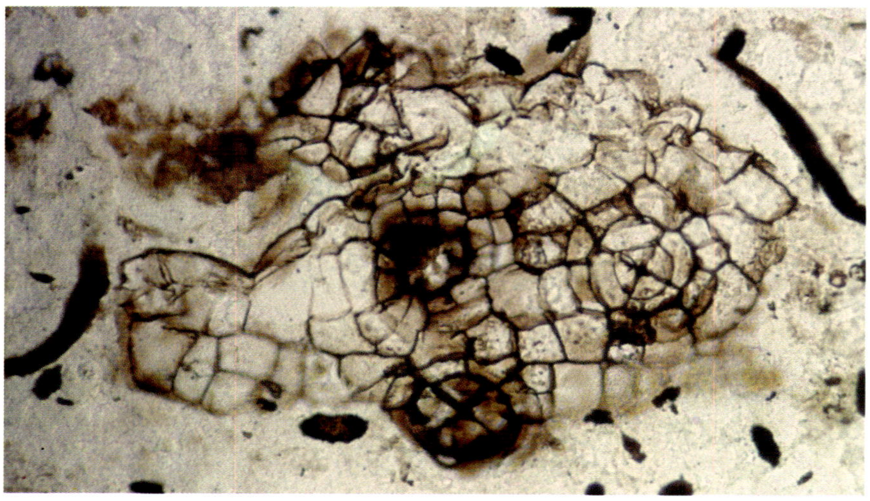

2 *Lepidodendron esnostense*
archegonia on the surface of
megagametophyte, 250 x

The organic material of the plant cell walls is then preserved. As a result, preparations of delicate structures may thus be obtained using appropriate techniques, e.g. etching by appropriate acids. This exceptional preservation is illustrated by a very fragile female gametophyte from this site germinating in a lycopsid megaspore of *Lepidodendron esnostense* which is only one millimetre in diameter.
The preparation allows several female reproductive organs, called archegonia to be studied from both surface and longitudinal views (Figs. 1-2), and thus demonstrates the reproduction process of these ancient plants was similar to that of modern *Selaginella* and *Isoetes*.

The Esnost flora was dominated by ferns with *Botryopteris antiqua* being the most common element. This plant had a scrambling growth habit.
Its delicate leaves were perfectly preserved, including the fragile phloem tissue (Fig. 3).
A reconstruction of the palaeo-communities at Esnost by Rex

distinguished a peat forming area colonized by *Archaeocalamites* (Fig. 4) and another area with the dominant fern *Botryopteris* fossilised in situ. In the latter case, evidence of plant-animal interaction was found inside the *Botryopteris* rachises with borings containing coprolites composed of sporangia and spores from the same fern.

Another chert-litholgy, dominated by the robust zygopterid fern *Diplolabis* (Fig. 5), was characterized by a variation in the preservation of the organic material. The material was sometimes very faint or absent, having been replaced by yellow silica. In this case the mechanism of mineralization was comparable to that of silicification of most petrified woods.

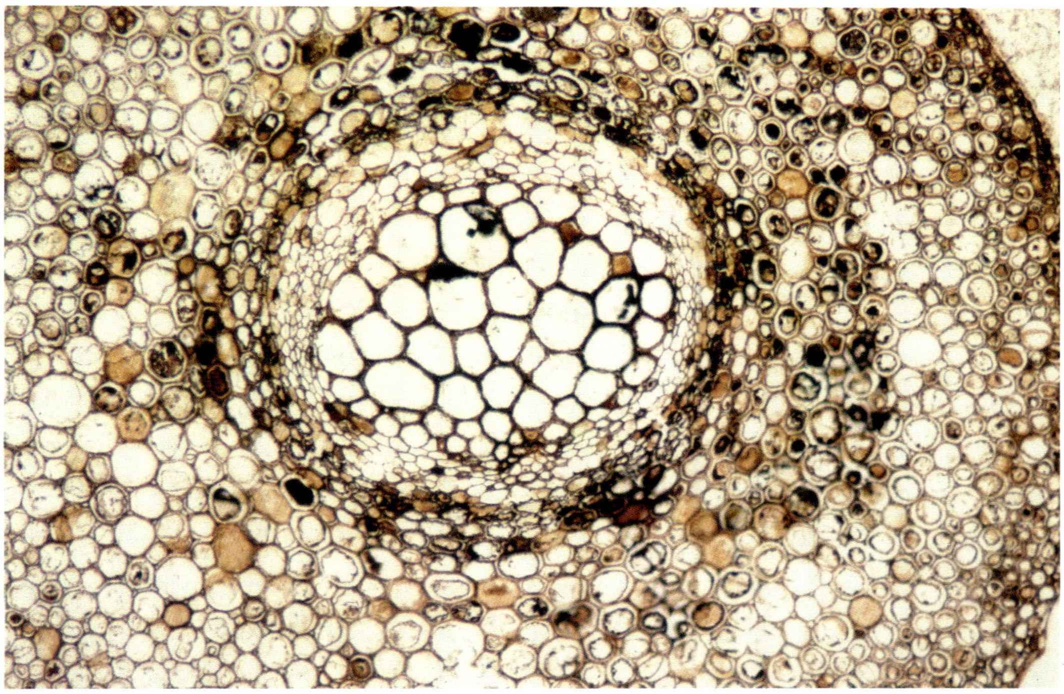

3 *Botryopteris antiqua* fern leaf rachis with all tissues preserved 100 x

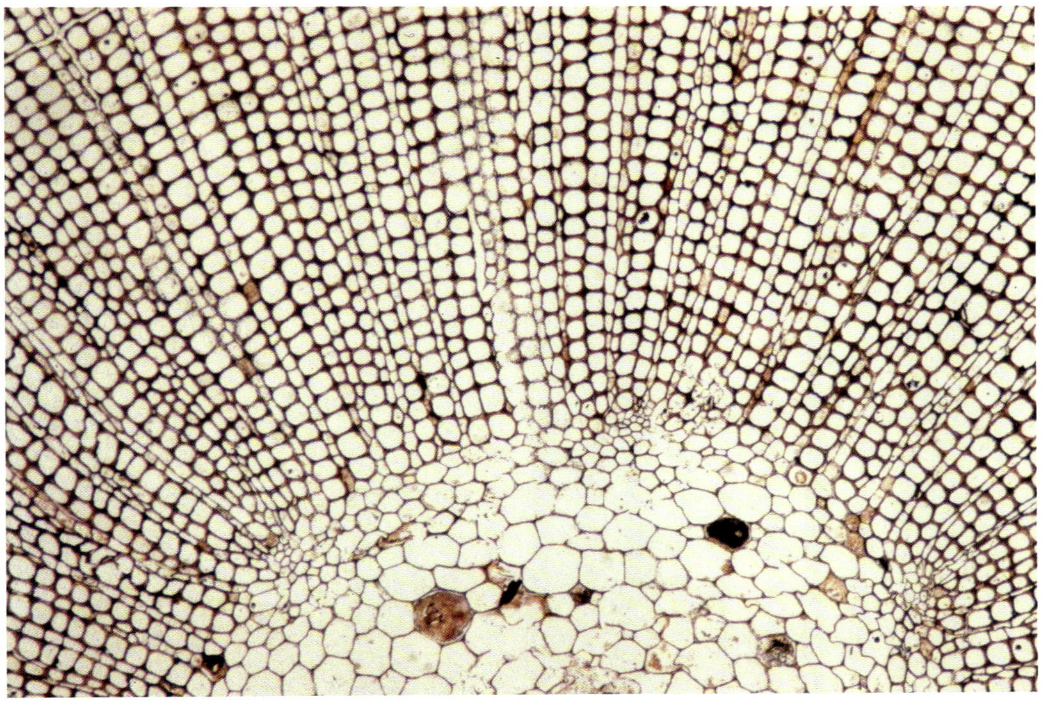

4 *Archaeocalamites* root Detail of pith and wood 50 x

In Scotland there are several floras of similar Viséan age also associated with volcanic settings. The most famous locality is Pettycur, along the shore of Firth of Forth, north of Edinburgh where fossil plants were first described by Williamson from 1871 to 1895.
Here the plant deposits are associated with a series of basaltic lavas, therefore the plants differ chemically, their permineralizing mineral being calcite instead of silica.
The Pettycur and Esnost floras are similar with several important species in common but their palaeoenvironments were different. Pettycur limestone was a calcareous peat similar in aspect to that of coal ball (Rothwell, in this volume).

The fossilpart contains abundant lycopod remains but also *Archaeocalamites* and the fern *Botryopteris antiqua* as in France. At Pettycur the fern *Diplolabis* occupied a separate growth site, but it is always found with its close relative *Metaclepsydropsis* (Fig. 6).

Another Lower Carboniferous locality in Scotland is at East Kirkton, 30 km west of Edinburgh. This locality has been known since the 19th century as a volcanogenic sequence showing hot-spring features where occasional fossils including plant compressions, were found.
Recently the site attracted considerable attention leading to the discovery of the oldest fauna of fully

5 *Diplolabis*, fern leaf with X-shaped xylem 15 x

6 *Metaclepsydropsis* fern leaf rachis, 10 x

7 Outcrop of East-Kirkton (Scotland) in 1992 type section of the East-Kirkton limestone horizon, which yield a famous terrestrial biota of late Viséan age

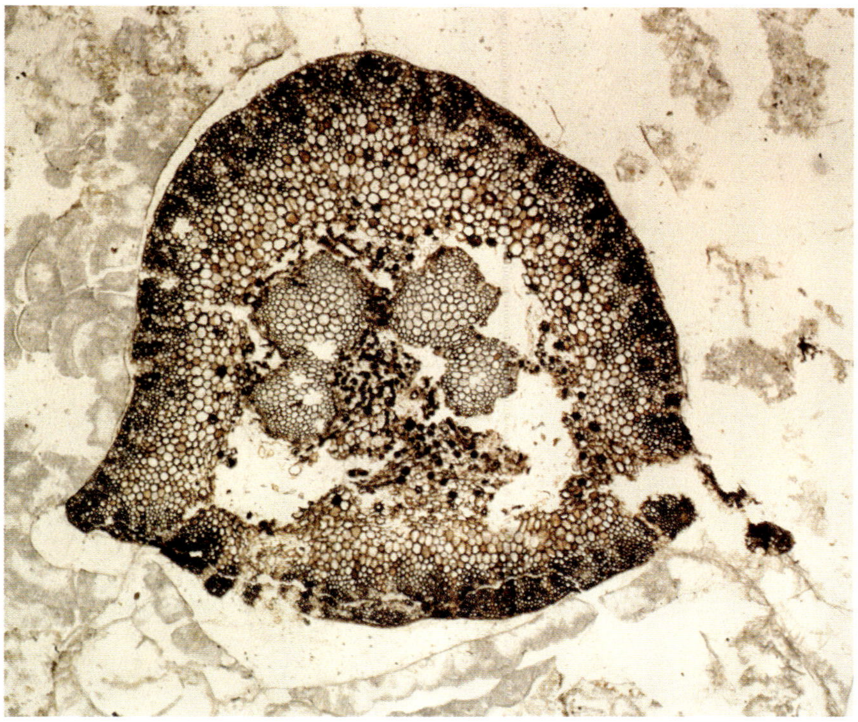

8 *Lyginorachis* seed plant leaf, 25 x

terrestrial tetrapods. A bed-by-bed excavation of the site allowed a multidisciplinary investigation of this unique terrestrial community which is approximately 335 millions years old. This was also an opportunity to recognize several fossil plant assemblages (Fig. 7).
Of particular interest are the anatomically preserved plants occurring in laminated limestones or in stromatolitic nodules which are without equivalent in other volcanic environments.

The plants are exceptionally well preserved as demonstrated by *Lyginorachis* (Fig. 8), the petiole of a seed fern that had been growing at this site and possibly fossilized by hot-spring water.

The above examples of plants of Lower Carboniferous age emphasize the extraordinary significance of volcanic deposits that provide the minerals for the exquisite preservation of the delicate plant tissues of this age.

JEAN GALTIER

Pitus, ein Riesenbaum aus der frühen Steinkohlenzeit

Pitus, a Giant Tree of the Early Carboniferous Time

Plant fossils from the Lower Carboniferous of the Midland Valley of Scotland are unusually diverse and many are of particular interest because their internal structure is preserved. One fossil plant worth mentioning is *Pitus*, which is one of the most important fossil trees of Carboniferous age. This plant is of particular historical interest in connection with the classic works of Witham in 1830 and 1833, who illustrated for the first time the tissue anatomy of a fossil plant. This was made possible by the use of petrological thin sections prepared by William Nicol, starting with a thin slice of the petrified wood which was subsequently ground until it became translucent. Witham compared the anatomy of present trees with that of a few fossils; among them were the *Pitus* trees, which were both the oldest and the largest known at that time. The original *Pitus* trunks were more than ten meters long and more than one meter in diameter. They were found in the Craigleith quarry near Edinburgh and initially called "Craigleith fossil trees" by Witham. Later the plant was named *Pitus withamii*. Several trunks of *P. withamii* have been removed from this site and two are still displayed, one in the Royal Botanical Garden in Edinburgh (Fig.1), and the other in front of the National Museum of Natural History in London.

Many other trunks of *Pitus* were discovered at this time at several localities in Scotland. These were also of Lower Carboniferous age. Several of these trunks illustrated by Witham, who described them under the names of *Pitus antiqua* or "Lennel Braes tree" and of *Pitus primaeva* or "Tweed-Mill tree". Witham illustrated not only the wood but also the central region of

1 *Pitus withamii* trunk displayed in the Royal Botanical Garden, Edinburgh

the trunk, or pith, which is composed of a soft tissue. Nicol's original thin sections, which were illustrated by Witham, are housed in several institutions, in particular the collections of the National Museum in London and in Glasgow. These old thin sections are still of excellent quality and can be compared with the original illustrations of Witham.
As an example, a transverse section of the wood of *Pitus antiqua* (Fig. 2) shows the regular radiate arrangement of the thick-walled wood cells, known as tracheids, with the intercalated rays of dark thin-walled cells. The same tissue observed in a tangential section (Fig. 3) shows that some tracheids are very long and has narrow rays. This contrasts with the wood of the related species, *Pitus primaeva* (Fig. 4) which possessed very broad rays.
A radial section (Fig. 5) reveals the numerous minute openings, called pitting, in the tracheid walls. Information provided by the observation of these sections demonstrated that different types of *Pitus* trees existed during most of the Early Carboniferous time, 350 to 330 million years ago. Moreover, the widespread occurrence of these fossils in Scotland suggest that *Pitus* forests were common. A remarkable feature of these trees is their size. The largest specimen of *Pitus withamii*, initially found in the Craigleith quarry, had a maximum length of 17 meters and a basal diameter of 1.8 meter, although another specimen was more than 20 meters long. Recently, trunks of *Pitus primaeva* up to 2.4 meters in diameter have been described by Long, but personal observation of these trunks has revealed that they are incomplete and that their original

diameter was much larger. These dimensions surpass those known for other trees of Lower Carboniferous age. It is significant that *Pitus* was not only larger than *Archaeopteris-Callixylon*, the biggest Devonian tree, but it was certainly also larger than any of the well-known younger Carboniferous coal swamp trees (like *Lepidodendron*, *Calamites, Cordaites*).

Pitus was a giant tree that became extinct at the end of the Lower Carboniferous. It was not surpassed in size until the rise of large cordaites and conifers with *Dadoxylon* type wood well represented in the Lower Permian of Chemnitz, Autun and other areas of the world.

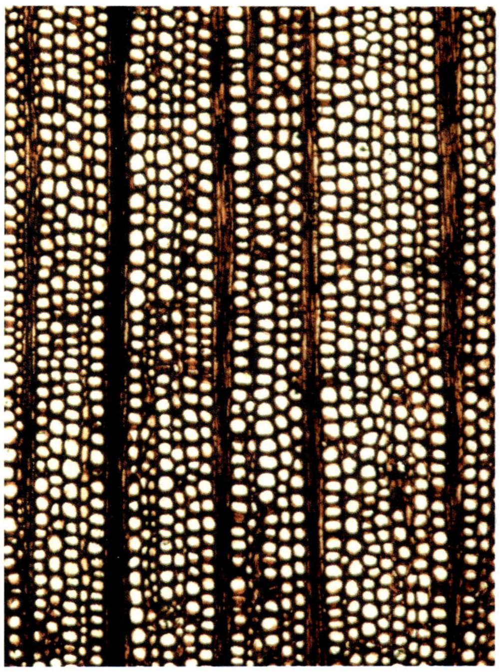

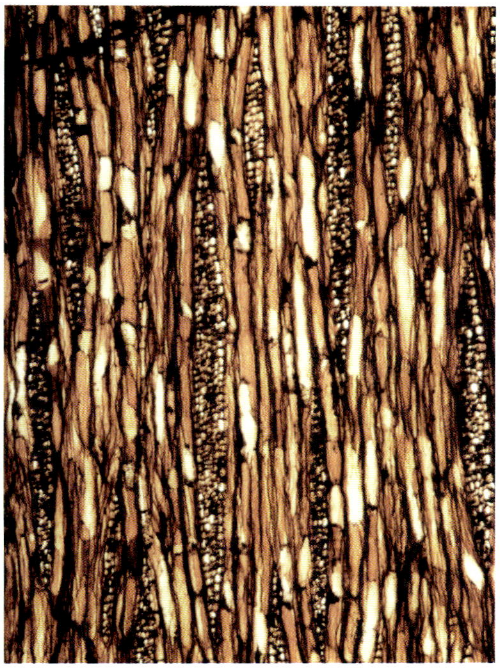

left:
2 *Pitus antiqua* wood, transverse section
35 x

right:
3 *Pitus antiqua* wood, tangential section
35 x

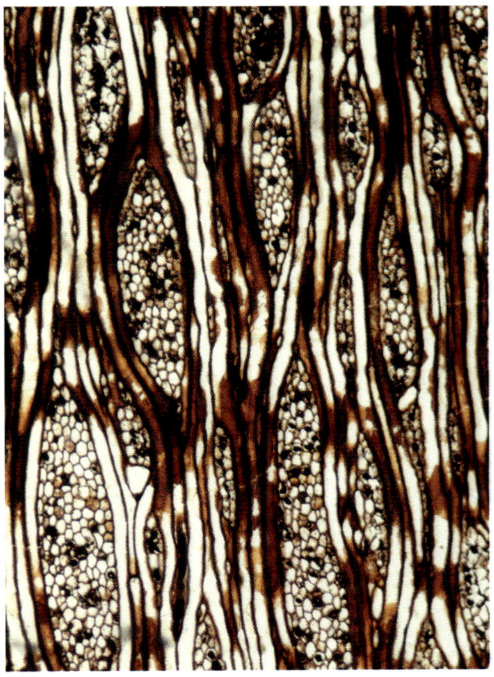

left:
4 *Pitus primaeva* wood tangential section showing very large rays
35 x

right:
5 *Pitus withamii* wood radial section of the Craigleith tree
70 x

GAR W. ROTHWELL

Coal Balls - Bemerkenswerte Zeugnisse paläozoischer Pflanzen und Pflanzengesellschaften

Coal Balls - Remarkable Evidence of Palaeozoic Plants and the Communities in Which They Grew

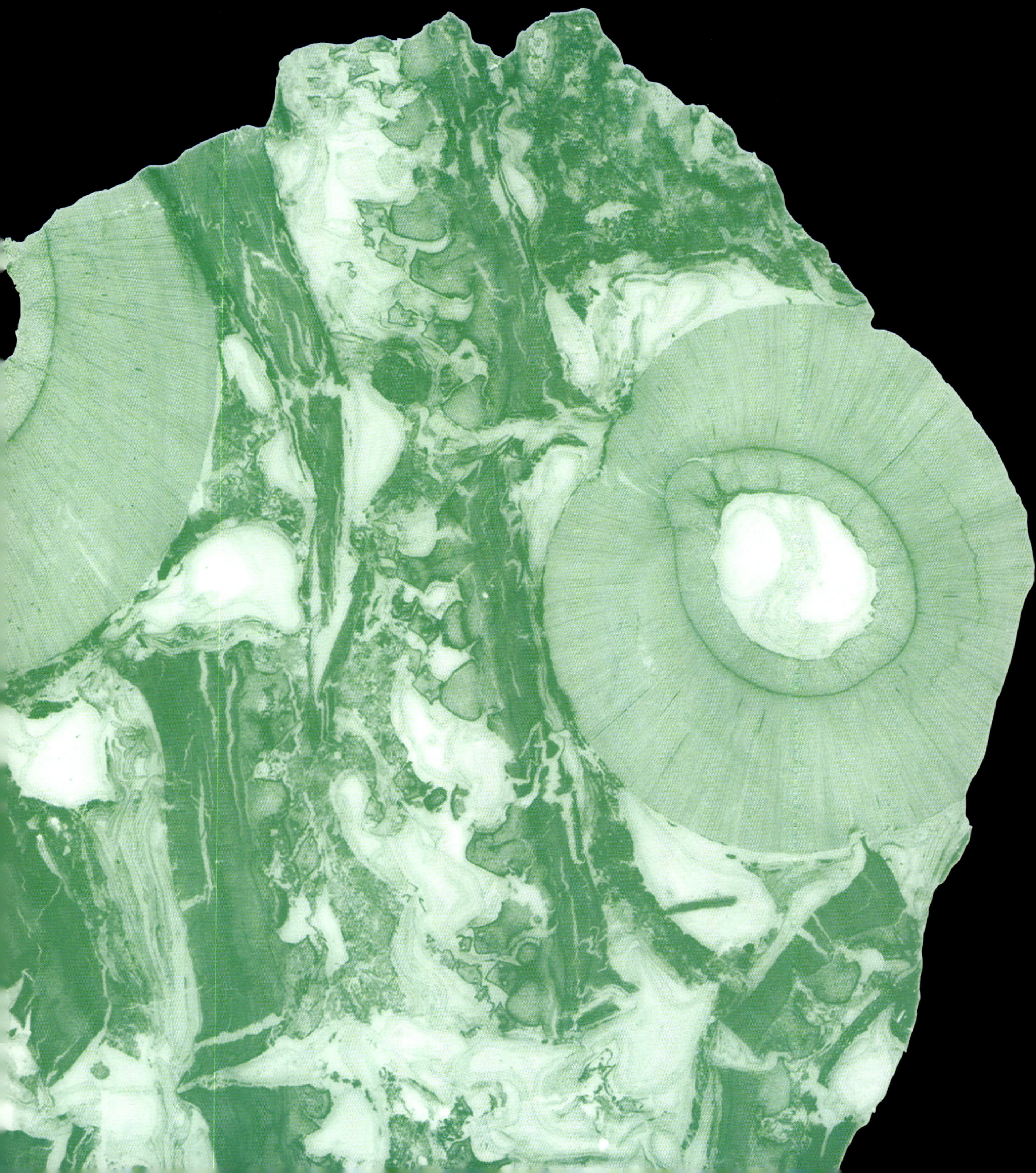

Thoroughout the history of life special types of geological processes have preserved remarkably complete evidence for many types of organisms. Because of a cellulose wall that surrounds each cell, plant fossils are particularly likely to display internal structures, with even the most delicate features preserved in some instances. Anatomical preservation is often limited to isolated plant parts, but nearly complete plants and even entire plant communities also can be preserved. Spectacular examples of such preservation occur in sediments that range from the Lower Devonian of Scotland through the Permian and Triassic of Antarctica, Jurassic of India and Eocene of North America, to the Miocene of North America and Japan. The combined biological, depositional and preservational conditions required to produce anatomically preserved floras are relatively rare, with different modes of preservation sometimes characterizing particular segments of the geological record.

The most abundant and thoroughly studied of all anatomically preserved fossil floras occur in calcareous concretions known as coal balls. Although the processes that led to the formation of coal balls are not fully understood, we do know that they are limited primarily to the equatorial tropics of the Pennsylvanian and Permian periods, and that they preserve the wetland vegetation that is the source of the most abundant Palaeozoic coal deposits. The earliest coal ball studies were initiated nearly 150 years ago in Europe, and by the 1920's Lower Pennsylvanian plants of England, Holland and Germany had been extensively studied. Abundant coal balls were recognized in the Urals of Russia early in the 20th century and in the United States in the 1930's. Together, these provide material for the most thorough characterizations of extinct vegetation thus far conducted from throughout the fossil record. More recently, coal balls have been discovered in both Lower and Upper Permian deposits of China, and these hold great promise for further extending our understanding of Paleozoic coal swamp vegetation. Coal balls derive their name from their occurrence in coal seams, and from the nearly spheroidal shape of the earliest examples found in 19th century colleries of England. Coal balls from other sources are often more irregular in shape (Fig. 1), and sometimes occur in large aggregates or even solid carbonate masses that occupy nearly the whole coal seam at certain places. Most coal balls are found in areas where swamp communities were periodically inundated by carbonate rich sea water. Accumulated debris on the floor of the swamp apparently became infiltrated with carbonates that solidified to encase the plant material before the rest of the deposit was compressed to form coal. Therefore, the plant material in coal balls provides an accurate picture of the coal-forming vegetation.

Palaeobotanical techniques for preparing coal balls have been refined through time, but thin sections have been used to reveal internal anatomy since the earliest studies. In modern studies the coal balls are cut into slabs, and cellulose acetate peel preparations, that produce a thin section of the plant material, are made from each cut surface (Fig. 1). Repeated peeling of coal ball surfaces allows for the serial thin sectioning of plant organs, and for the eventual reconstruction of entire plants.

Coal Ball Studies

Coal balls have provided material for many pioneering studies of plant fossils, including organ development, growth architecture, pollen and spore ultrastructure, reproductive biology, community structure, successional ecology, and vegetational turnover through time.

Almost all of the plants contained in coal balls are assignable to five major groups. These include lycophytes, equisetophytes, ferns, seed ferns, and cordaiteans.

1 Coal ball cut into slabs, one slab showing plant material on etched surface
Steubenville Locality, (Duquesne Coal, Conemaugh Group)
Upper Pennsylvanian of Ohio, 0.5 x

2 Peel showing two stems with flattened cortex and one cone (i.e., *Flemingites*; at arrow) of a *Paralycopodites* lycophyte tree in cross section. Lewis Creek Locality, Copland Coal, Magoffin Member, Breathitt Formation) Lower/Middle Pennsylvanian of Kentucky, original size

Lycophytes: The lycophytes are represented primarily by a diverse assemblage of plants that are most closely related to the living quillwort *Isoetes*, and that comprise the dominant canopy trees in the Lower and Middle Pennsylvanian. Near the end of the Middle Pennsylvanian the most prominent family of arborencent lycophytes, the Lepidodendraceae, became extinct, and isoetaleans have continued to decline up to the recent. Our concept of these trees has been developed primarily from a species segregated from *Lepidodendron* as the genus *Paralycopodites* (Fig. 2). *Paralycopodites* trees flourished from early in the Lower Carboniferous to near the end of the Middle Pennsylvanian, and are characterized by dichotomously branched woody stems with closely arranged grass-like leaves and bisporangiate cones of the genus *Flemingites* (Fig. 2 at arrow) that occur at the tips of the branches. The thick cortex of the stems consisted of loosely arranged tissue that often decayed rapidly and collapsed around the woody stele, and bore the leaves on raised leaf cushions (Fig. 2). Well preserved cones of *Paralycopodites* and closely related lycophytes have yielded evidence of spore ultrastructure, gametophyte development, embryos, and even chromosomes, and the studies of coal-ball species have played a major role in developing a comprehensive understanding of the pattern of phylogeny for this important clade of vascular plants.
The extensive rooting system of *Paralycopodites* and other Palaeozoic lycophyte trees is known as the genus *Stigmaria* (Fig. 3). Although the downward growth and branching of *Stigmaria* is remarkibly similar to the massive rooting systems of conifers and flowering plant trees, *Stigmaria* actually represents a shoot system that is modified for rooting. *Stigmaria* is arguably the most abundant of Upper Palaeozoic plant fossils, and this genus produced a large percentage of the biomass for many Lower and Middle Pennsylvanian coals.

Sphenopsids: Sphenopsids are the second major group of coal swamp plants that were common in the Palaeozoic, and subsequently declined in diversity. Although they are represented in the modern flora by only the herbaceous genus *Equisetum*, coal swamps contained both shrubs and vines of the genus

3 *Stigmaria*, the rooting organ of a lycophyte tree, showing characteristic surface pattern and internal anatomy on cut surface
Lewis Creek Locality, original size

Sphenophyllum, and woody *Calamites* trees (Fig. 4). *Calamites* is probably more closely related to *Equisetum*, and differs from the living genus primarily by the production of abundant wood (Fig. 4), by leaves that are larger and less fused together (Fig. 5) than those of *Equisetum*, and by features of the cones. Like living species of *Equisetum, Calamites* plants have shoots that grow from a single apical cell, and have spores with highly elaborate structures called elaters.

Ferns: A diverse assemblage of ferns including representatives of the Zygopteridales, Marattiales and Filicales is represented in coal balls. The marattiaceous tree fern *Psaronius* became the dominant canopy tree in many coal swamp forests after the extinction of the Lepidodendrales near the end of the Middle Pennsylvanian. *Psaronius* trees produced a trunk that consists of an unbranched stem surrounded by a mantle of prop roots (Fig. 6). This genus is characterized by several cycles of vascular tissue in the stem (Fig. 6), and displays a crown of large, highly dissected fronds. Spores were produced in two rows of fused sporangia, known as synangia, on the abaxial surface of the pinnules (Fig. 7).
True ferns of the Order Filicales typically are small plants that contribute only a tiny fraction of the biomass in Palaeozoic coals. Nevertheless, they are common components of coal ball floras with wide systematic diversity and several forms of growth. The most common genera are assignable to the Ankyropteridaceae, Botryopteridaceae, and Sermayaceae. Some are rhizomatous herbs with emergent fronds, while others are vines or epiphytes. Our understanding of plant community structure in the Pennsylvanian coal swamps has been significantly strengthened by evidence for plant interactions that is preserved within the prop roots on the trunk of *Psaronius*. The genus *Ankyropteris* (Fig. 8) is a common example of a vine that grew up the trunk of *Psaronius*, but several other ferns and the sphenopsid *Sphenophyllum* also were rooted on *Psaronius* trees.

4 Large woody stem of a *Calamites* tree, surrounded by other plant debris
Steubenville Locality, 0.5 x

The protostelic stems of *Ankyropteris* produced both fronds with an "H" or anchor-shaped frond trace and axillary branches (Fig. 8) that protruded from the surface of the *Psaronius* trunk.

Seed Ferns: The most diverse seed plants found in coal balls are assigned to the unnatural group commonly known as seed ferns. Three orders of seed ferns, Lyginopteridales, Medullosales, and Callistophytales are present in coal balls of North America, Europe and Russia, and a fourth order, the Gigantopteridales, occurs in Permian coal balls of China.

Many seed ferns, such as the famous *Lyginopteris oldhamia* from Europe (Fig. 9) and species of *Callistophyton*, were scrambling shrubs and/or vines, whereas some species of *Medullosa* were also small trees with stems that produced abundant wood (Fig. 10). Most medullosan plants produced highly dissected fronds that forked one or more times, but the fronds of a few other species may have been pinnately dissected throughout (Fig. 11).
Much of our understanding of the evolution of seed plant reproduction has been developed from plants

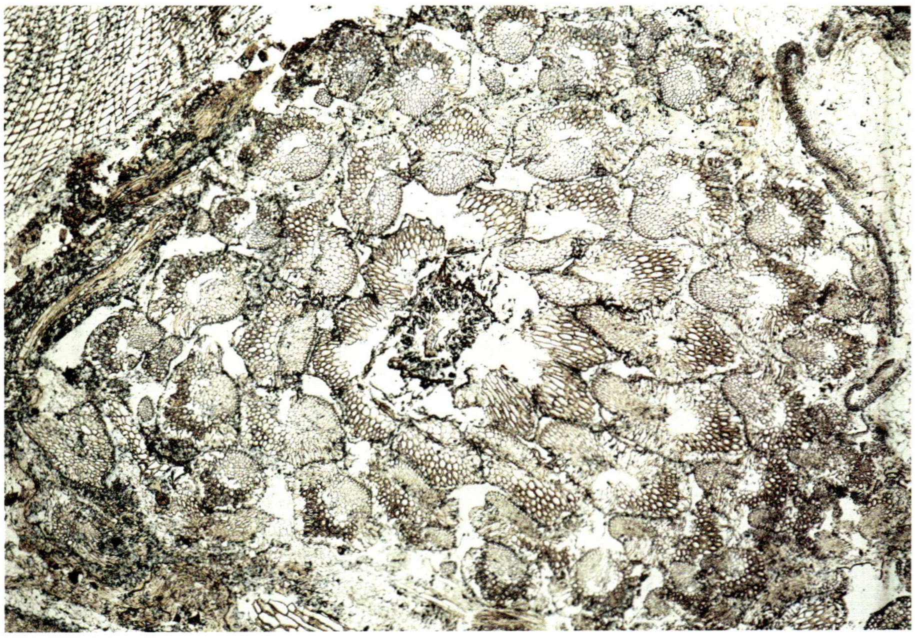

5 Cross section near the tip of a *Calamites* branch showing whorls of *Asterophyllites*-type leaves around tiny stem
West Mineral Locality (Fleming Coal, Cabaniss Subgroup, Cherokee Group, Demoinsean Series)
Middle Pennsylvanian of Kansas, 10 x

preserved in coal balls.
For example, developmental studies of the seeds and pollen organs of *Callistophyton* have shown that microgametophyte development was comparable to living conifers, that some Palaeozoic ovules were pollinated via a pollination drop at an early stage of development, that the pollen produced a pollen tube (Fig. 13), and that mature ovules had cellular megagametophytes, archegonia and eggs that are similar to those of living cycads, *Ginkgo*, conifers and *Ephedra* (Fig. 14).

Cordaiteans: Cordaitean gymnosperms were the largest seed plants preserved in both Pennsylvanian age coal balls of Euramerica and in the Permian coal swamps of China. Although some species grew as shrubs, many other plants formed trees of varying sizes and growth architectures. Cordaiteans produced highly specialized, compound cones that each produced bore either pollen or ovules. Euramerican species of cordaiteans are assignable to either *Cordaixylon* or *Mesoxylon*, whereas additional diversity is present in Permian coal balls of China.

6 Cross section of *Psaronius* tree fern, at a relatively high level in trunk, showing large stem surrounded by a relatively thin zone of prop roots
Calhoun Locality (Calhoun Coal, McCleansboro Group)
Upper Pennsylvanian of Illinois, 0.5 x

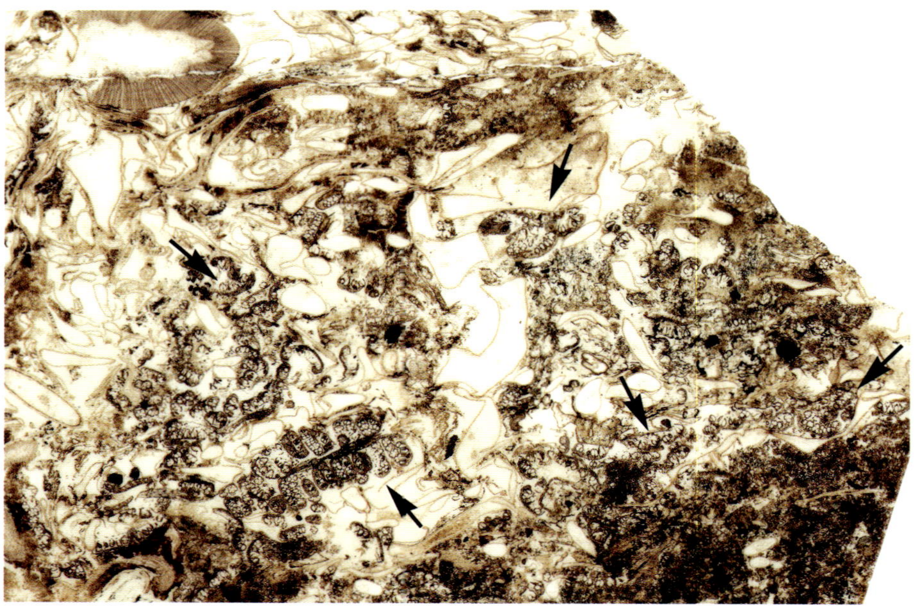

7 Fertile frond parts of a *Psaronius* tree showing pinnules with synangia of the genus *Scolecopteris* (arrows)
Steubenville Locality, 2 x

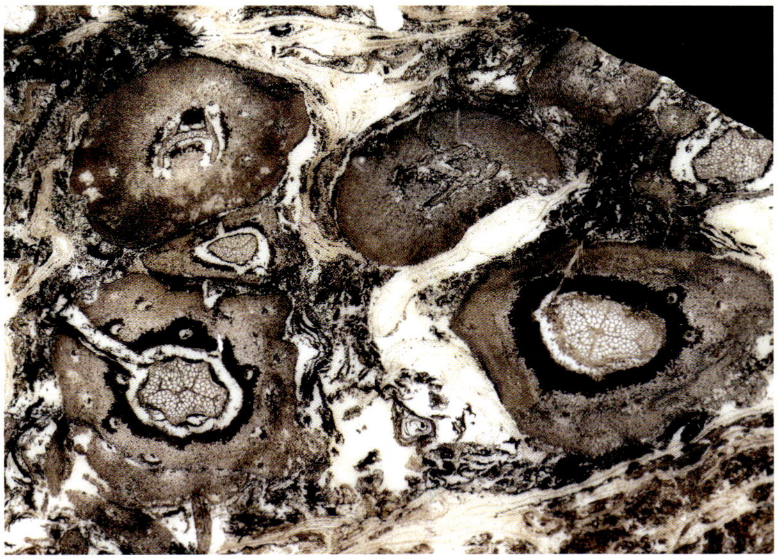

8 Cross sections of stems, and frond bases of the true fern *Ankyropteris* that are growing as a vine on the surface of a *Psaronius* tree fern trunk, Sahara Locality (Herrin No. 6 Coal, Carbondale Formation, Kawanee Group) Middle Pennsylvanian of Illinois, 3 x

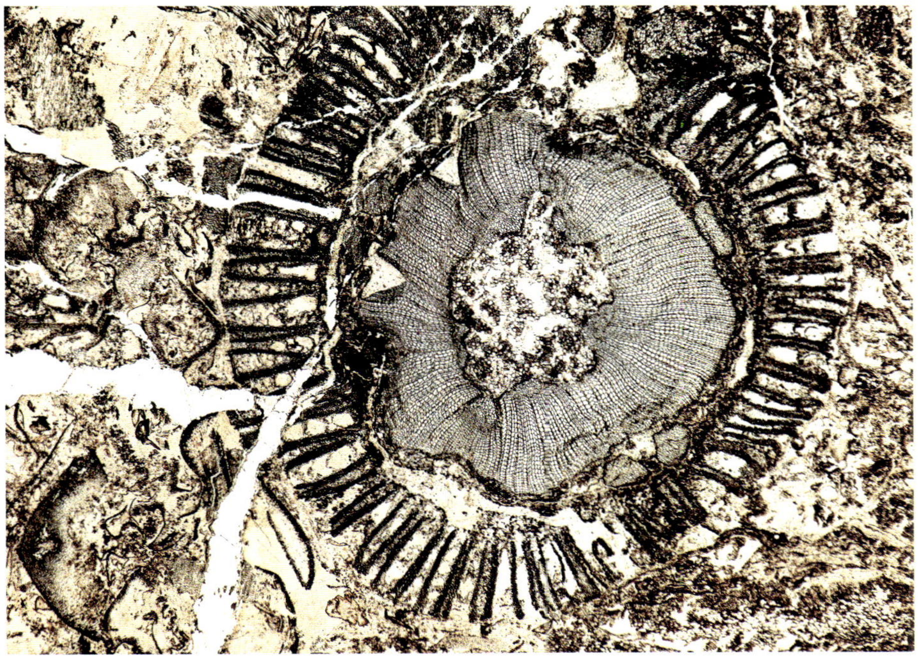

9 The famous seed fern *Lyginopteris* showing a woody stem with the diagnostic cortex that looks similar to the "roman numerals" on a clock face, Burnley Locality (Union Seam, Langsettian), Lower Pennsylvanian of Lancashire, England, 3 x

Mesoxylon priapii from the Upper Pennsylvanian of Ohio shows the representative woody stems and wide, strap-shaped leaves that characterize cordaiteans as a whole (Fig. 15). Cordaitean plants are more closely related to modern conifers than is any of the living groups of seed plants, and coal ball species of cordaiteans have contributed greatly to our understanding of reproductive biology and phylogenetic relationships among coniferophytic gymnosperms.

Ecology and Evolution

Because of the in situ preservation of plant remains, coal balls comprise an ideal source of palaeoecological data for interpreting vegetational patterns and testing evolutionary models. The pioneering paleocological studies of Phillips and others employ quantitative data that is derived largely from the vegetational analysis of coal balls. This work establishes that the general composition of Pennsylvanian wetland floras includes the five major groups described above, and it further

10 Coal ball that consists entirely of the stem of the seed fern genus *Medullosa*. Berryville Locality (Calhoun Coal, McCleansboro Group) Upper Pennsylvanian of Illinois nearly original size

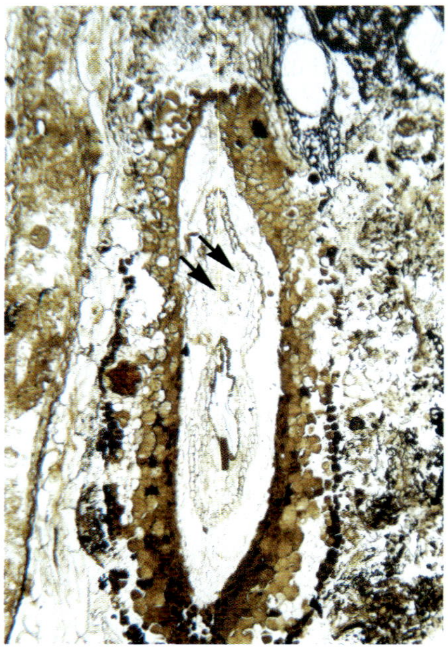

12 Immature ovule of the seed fern *Callistophyton,* two pollen grains (arrows) in the large pollen chamber Sahara Locality, 60 x

11 Split surface of coal ball showing a large medullosan seed fern frond of the genus *Myeloxylon,* Steubenville Locality, 0.5 x

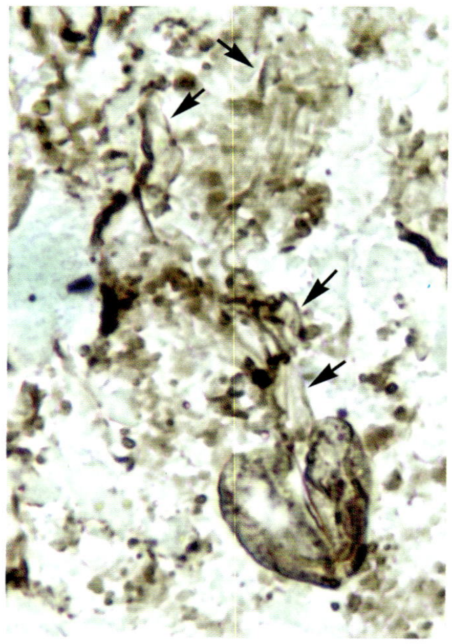

13 Close up of one pollen grain from the immature ovule in figure 12, showing branched pollen tube (arrows), 800 x

demonstrates that different species characterize different types of swamp environments at different times and places during the Pennsylvanian. These studies document extinction of the Lepidodendraceae and Diaphorodendraceae near the end of the Middle Pennsylvanian, and the rise to dominance of psaroniaceous tree ferns in the Upper Pennsylvanian. They convincingly correlate vegetational changes with global climatic patterns that are driven largely by periodic fluctuations in Southern Hemisphere glaciation. Such broad scale applications of coal ball data paint a dramatic picture of complex Palaeozoic ecosystems, that respond to environmental perturbations derived from regional climatic changes associated with global warming. In this regard, coal ball data provide a valuable analogue for predicting the nature of modern floristic changes that are expected to result from the ongoing global warming of today.

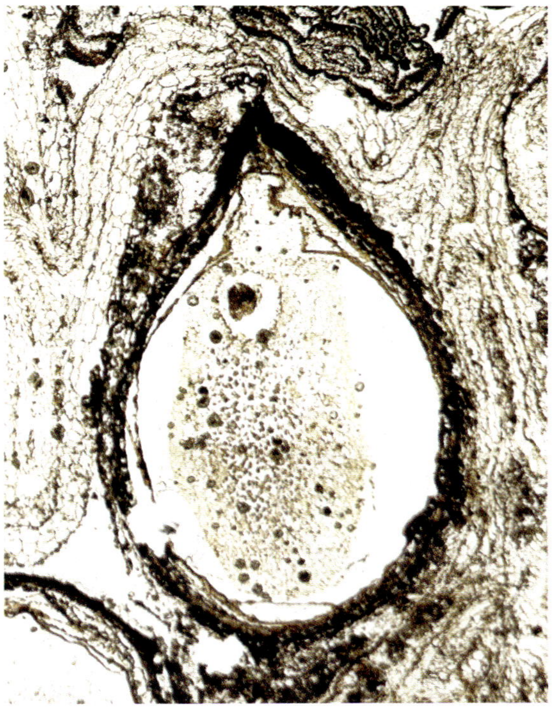

14 Mature ovule of the same species shown in figure 12, showing large cellular megagametophyte with an archegonium and egg, Sahara Localtiy, 40 x

15 Stems and leaves of the cordaitean tree *Mesoxylon* Steubenville Locality, original size

16 Coal ball peel showing accumulated plant debris, including a *Psaronius* tree fern trunk (arrow A) with *Ankyropteris* fern vines (arrows B), rootlets of *Stigmaria* (arrow C), *Lepidocarpon* megasporophylls (arrows D), and other genera of coal swamp plants Sahara Locality, original size

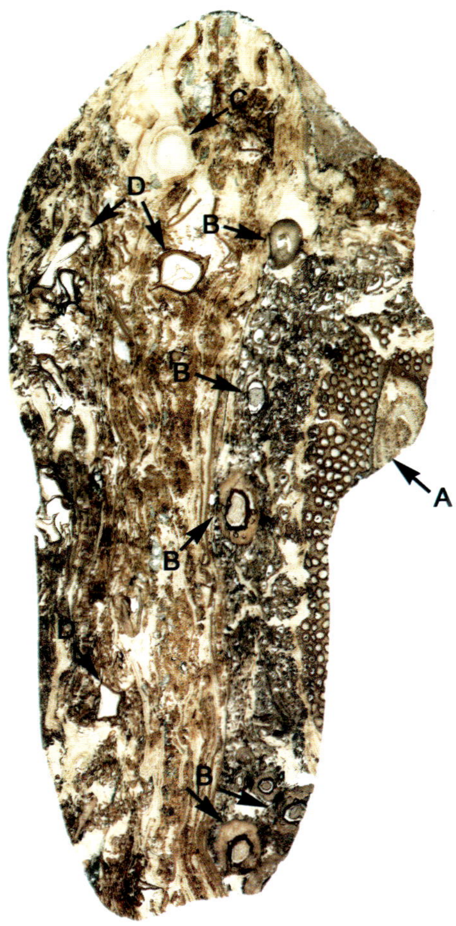

RONNY RÖßLER

Die steinernen Bäume vom Kyffhäuser, Deutschland

The Petrified Trees From the Kyffhäuser, Germany

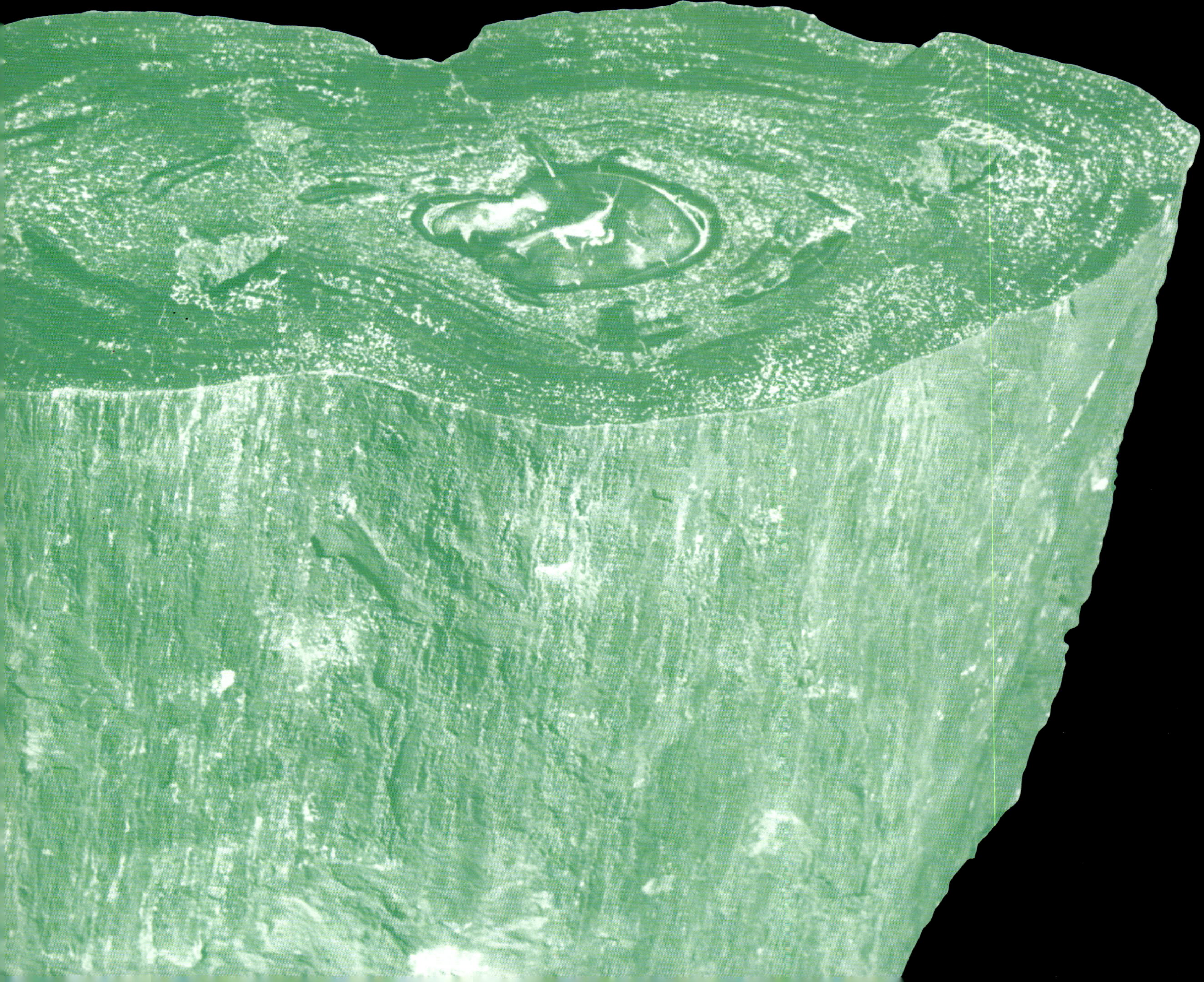

Old castles and palaces have tales to tell. According to local legend, it was here that Emperor Barbarossa and his courtly retainers withdrew into underground galleries beneath the Kyffhäuser massif. Once each century, the fabled emperor supposedly awakens from his sleep and sends a youth up to the surface to see whether the ravens are still flying. If the boy reports that the birds are indeed still soaring above the peaks, the bearded regent must continue his hibernation. It now seems as though the slumbering emperor can continue to dream undisturbed because during the past twenty years the ravens have returned to the old walls that rise atop the Kyffhäuser's slopes.
Located in central Germany, the Kyffhäuser is a small, wooded, mountain massif measuring roughly 10 km along one side and 5 km along the other side. A monument (Fig. 1) was built between 1890 and 1896 and erected atop its summit. The peak rises to an elevation of 477 m a.s.l. and offers a splendid view over the geomorphologically highly diverse and very interesting "Golden Plain" landscape. Caves, landslides, dry valleys, subterranean water-courses, and, petrified trees are only a few of the fascinating natural features found in this scenic region in northern Thuringia.

Known for Centuries, Yet Nonetheless Still Fraught With Enigmas
We're particularly interested in the Kyffhäuser because it's the site of Germany's geologically oldest petrified forest. The earliest geological and palaeontological literature (e.g. texts by J.E.I. Walch, J.F.W. Charpentier, J.C. W. Voigt, and J.C. Freiesleben) already mention silicified woods from the Kyffhäuser. Bernhard Cotta's 1832 description of various psaronias was likewise based on a specimen from Kyffhäuser (Figs. 2-3). But Cotta was wrong as far as his attempt at its identification was concerned: No genuine psaronias have ever been identified at this site, although the flecked and spotted coloration of

1 Kyffhäuser monument
Numerous Dadoxyla are contained within the conglomerate sandstone strata at the feet of Emperor Barbarossa.

2 Complete cross-section of a Kyffhäuser silicified wood; the original of Cotta (1832)
Geological Institute of the Freiberg University of Mining and Technology
diameter 50 cm, photo: Noll

these silicified woods frequently misled the unwary and caused them to confuse these specimens with the unmistakably speckled appearance of the *Psaronius*. This speckling in the Kyffhäuser specimens is an inorganic and only apparently morphological structure which prompted earlier researchers to give these woods the German name "Punctsteine," which translates as "speckled" or "spotted" stones. Unger used this appellation in 1847 to describe comparable pieces of silicified wood from the Elbe gravels which he placed among woods from conifer trees. Subsequent collectors likewise erred when they surmised that these same woods in fact were possibly derived from ancient palm trees. Göppert, "the father of German palaeobotany," correctly recognized their gymnosperm nature and assigned them to *Araucarites schrollianus*, which was assumed to be conifer wood. Göppert wrote that these fossilized woods were used in road construction between Kelbra and Bad Frankenhausen. These specimens were already too avidly coveted in those years for them to have been wasted in mere roadbeds. Göppert confirms their avidly sought status when he laments that "... only very little remains of the formerly rich abundance whose praises have been sung for more than a century...." He further notes that "I found only two trunks in their natural depository; and of the other trunks which doubtlessly existed here in the past and which seem to have been gathered as souvenirs, I found only a number of trunk fragments which were haphazardly strewn around an old linden tree at the beginning of a street behind the town and in a nearby wayside ditch, where they lay at the disposal of anyone who wanted them, a situation to which their numerous [recent] marks and traces testify." The literature records that trunks were found here which measured as much as 20 meters in length and with diameters of nearly 1.5 m.

300 Million Years Ago

The Kyffhäuser massif rests atop an old basement of granite and gneiss which formed during the Variscic. Above this base lie extensive packets of reddish colored, sandy to pebbly, riverine alluvia. These materials are products of the erosion of the Mid-German Crystalline Rise and were deposited on one another, depending upon variations in the transport energy of the palaeo-river system. Their deposition occurred during the latter part of the Upper Carboniferous, nearly 300 million years ago. This was the same era during which true psaronias comprised one portion of the abundant vegetation that flourished near here (at Wettin, north of Halle) and laid their biomass into the peat moors which formed the basis for the genesis of coal seams. Based on examination of the structure of the sediments and the attributes of the strata of the sandstone bodies, geologists surmise that the drainage direction of these ancient rivers ran from SW to NE. These sandstones are rich in kaolinized feldspar and occur in layers up to 500 m thick and include cross-bedded material. Coarse-

3 Fragment of a specimen of silicified wood from the Kyffhäuser area showing the typical flecks which formerly misled scientists to confuse such specimens with psaronias; the original of Cotta (1832): *Psaronius helmintolithus*
Museum für Naturkunde Berlin, 0.5 x

4 Former small sandstone quarry at the Kyffhäuser Massif showing horizontally embedded woods

5 Flattened silicified woods in the Upper Carboniferous sandstone beds of the Kyffhäuser Massif

6 Microphotograph of a concentrically structured fleck in wood Etzrodt collection, Borxleben, 14 x, photo: Noll

grained strata that often contain large silicified trunks (Figs. 4-5) have been interpreted as the remains of pebble banks. It is conjectured that these trunks were uprooted by strong rains, floated downstream in the swollen rivers, and finally deposited as driftwood on the pebble banks. On the same level as these sandstone sets, silicified woods can be found up to the Rothenburg region, north of Halle, where they are repeatedly unearthed in smaller sandstone quarries of the Siebigerode Formation.
Examination of these silicified woods suggests that the coarser alluvial deposits at Kyffhäuser in which the woods are found would seem to be the proximal equivalents of the Siebigerode sandstones. Unfortunately, we still do not know where these trees originally grew, although it seems likely that their forests once stood somewhere to the southwest of the Kyffhäuser. The finely stratified accompanying deposits of sandstones, as well as siltstones and shales from small ponds which formed on the floodplains, frequently contain plant detritus. In addition to this, albeit rarely, the deposits have also yielded fossilized arthropod trackways and insect remains. Carbonate-rich fossil soils suggest the existance of semiarid climatic conditions towards the end of the Carboniferous.
Later, approximately 258 million years ago, the Kyffhäuser became a small island whose shores were lapped by the surf of the Zechstein Sea. Deposits which formed range from the copperslate to erosion-prone deposits of carbonate, gypsum, and salt from the Upper Permian. These are responsible for subsequently formed karst structures such as dolines, caves, sinks, or barrows which currently characterize the appearance of a typical karst landscape.

Botanical Nature and Petrifaction – Still Mysterious Today ?

In addition to the quandary surrounding the still-unknown locations of the original forests, the petrified trunks of these trees also raise a number of unanswered questions. These queries involve, on the one hand, the botanical nature and taxonomic status of the plants themselves and, conversely, the details of the processes which led to their fossilization.

7 "Wood crystals" inside a specimen of silicified wood from the Kyffhäuser The quartz pyramids developed within the intact cells. Etzrodt collection, Borxleben nearly 20 x, photo: Noll

The first complex of questions has often been the subject of scientific research. Unfortunately, the often inadequate preservation of cellular details poses a nearly insurmountable obstacle to students of the anatomy of these woods. The most recent investigations were conducted by Karl Mägdefrau in 1958. All of the thin sections which he studied yielded one and the same consistent message: the silicified woods from Kyffhäuser are all derived from *Dadoxylon schrollianum*, a synonym for *D. saxonicum*. That these specimens are conifer wood is suggested by anatomical characters, like the one or two rows of bordered pits in the tracheids, the primarily one-rowed rays, and the wedge-shaped primary xylem groups which protrude into the pith. Branching is unfortunately seldom present, even on longer segments of trunks. Among his notes, Mägdefrau mentions two adjacent branch departures which suggest a whorl; but he also writes about individual departing branches, which suggests that the Kyffhäuser woods were only sporadically, irregularly branched.
Finally, let's turn our attention to certain fossilization phenomena, like the pale, concentrically structured flecks found in the dark reddish-brown silicified wood. Recent investigations of Weiß provide some evidence that has led us to believe that they were created during an early stage of the mineralization process and could well represent the beginning of the silicification of the woody tissue (Fig. 6). If this were indeed the case, then the centers of the flecks (usually individual tracheids) could be interpreted as being the sites where the initial silicification first commenced. These flecks, which are very well preserved in this area of the wood's cells, may have stabilized the plant tissue by virtue of their early silicification; elsewhere in the specimen, the silicifying tissue would still have been subjected to deformations. There are numerous indications of swelling and shrinking; the latter probably correlate with occasional periods when the wood dried out. This would provide support for a theory that indicates the woods had already undergone their final embedding prior to their silicification. Many of the trunks found here show obvious traces of compaction. Perhaps the decaying kaolin-rich feldspars of the arkose-like sandstones were the silica source necessary for the woods' mineralization. Another special feature is the repeatedly observed growth of idiomorphic quartz pyramids within the intact cells (Fig. 7). Older opinions have suggested that quartz crystals formed only in hollows in previously decayed wood or as a result of Recrystallization. These would be worthy of reconsideration. Additional investigations will be necessary brand the remaining riddles surrounding the silicified woods from Kyffhäuser can be solved.

RONNY RÖßLER

Psaronienwälder am Fuße der Vulkane

Psaronius Forests at the Feet of Volcanoes

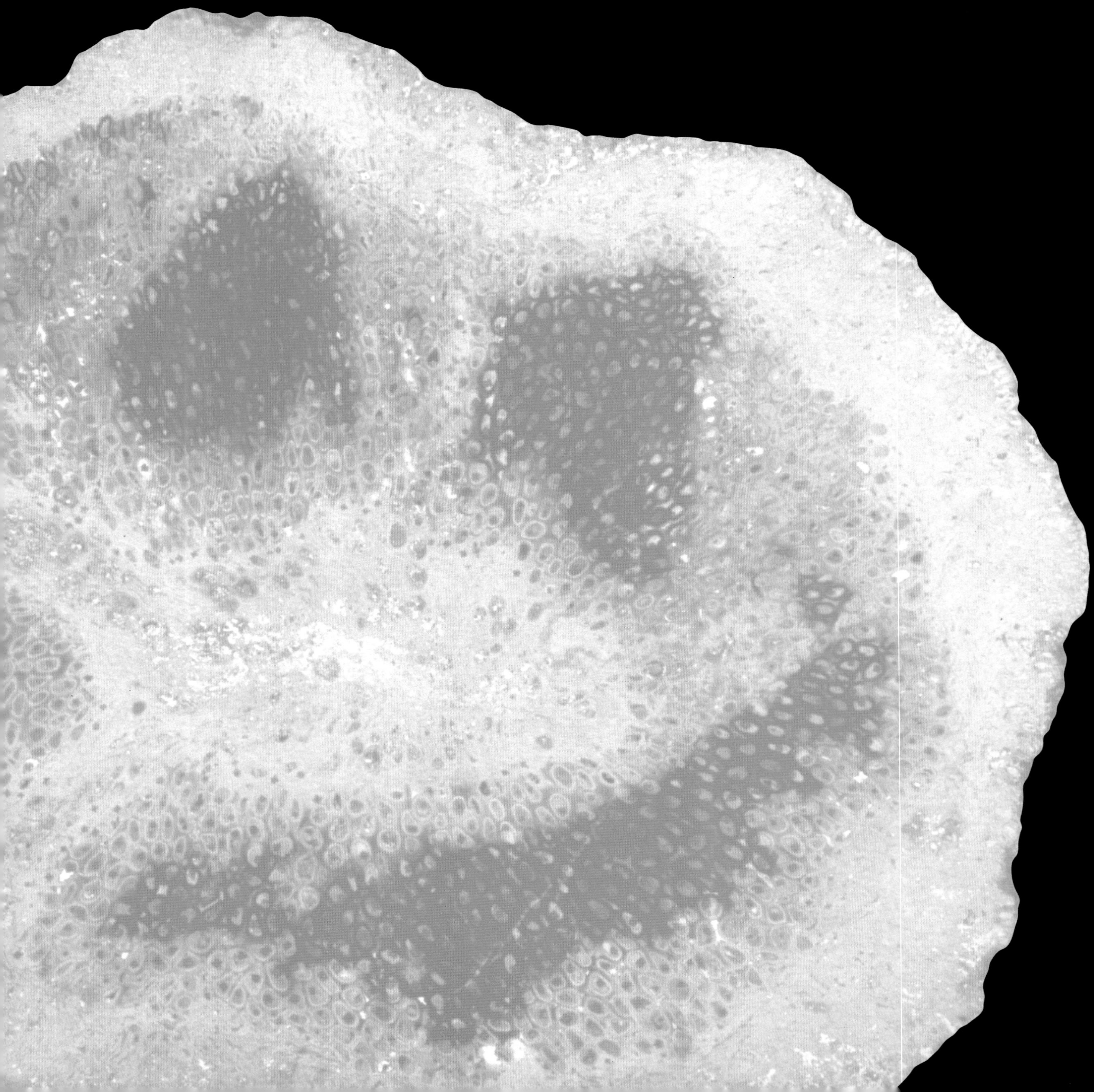

The fossilized forests of the European Upper Carboniferous and Lower Permian yield so many fascinating specimens from millions of years of the Earth's history that they fully deserve their outstanding palaeontological status. On the one hand, the history of their research coincides with the earliest days of paleobotany, and we can thank the first scientific investigators of Permian woods for coining such mellifluous names as "*Psaronius*" or "*Medullosa*." Nowadays, these genera are regarded as genuine classics, and specimens of them shouldn't be lacking from any collection. Valuable historical collections such as those of Tharandt forester H. Cotta, Breslau university professor H.R. Göppert, French researcher B. Renault, or Chemnitz teacher J.T. Sterzel contributed indispensably to the initial studies of palaeobotany. These collections still contain unique, museum-worthy legacies from the initial phase of this branch of science. Conversely, the majority of these Permian treasures would not have been preserved without the volcanism and its effects on the ancient ecosystems. Explosive volcanism isn't only a destructive and violent power, but can also create unique conditions which have led to the preservation of long-extinct forms of life. Nearly 290 million years ago, numerous volcanoes were active throughout Europe. Their peaks lined the edges of the colliding tectonic plates that were then forming the supercontinent of Pangaea. Ashes precipitating in the aftermath of their eruptions became beds in which coveted silicified woods were preserved. Many finds came from the foothills of the Riesengebirge (Nová Paka), Saxony (Chemnitz, Freital, Gnandstein), Thuringia (Manebach), and SW Germany (Palatinate), up to the Massif Central (Autun) in France.

In the following paragraphs, I'd like to explain, first of all, why these first studies of palaeobotany continue to make new insights possible, and to further show why they still confront contemporary researchers with so many unsolved riddles. As we shall see, the most puzzling items are oftentimes the classic specimens,

1 „Petrified Forest" of Chemnitz in front of the Museum für Naturkunde in Chemnitz, photo: R. Bartel

2 Detail of the stele of a distich *Psaronius*-tree fern collection of the Museum für Naturkunde Chemnitz, original size

stones which have been "turned over and over by generations of learned hands." When these frequently examined items are compared with newly discovered specimens and subjected to the closer scrutiny using the latest methods of scientific examination, these well-known older specimens disclose very new insights.

What Makes a Specimen of Silicified Wood Famous ?

As part of the testimony about the evolution of life on Earth, the "petrified forests" deserve to be included in the cultural legacy of humankind.
To preserve and use this irreplaceable inheritance wisely is a task for present and future generations of researchers. The fame of our fossilized treasures has increased thanks to the research and museum work of our predecessors. It was these learned individuals who first subjected the fossils to examination, put them on public display, and then explained them in a manner that could be understood by interested laymen. Let us therefore transport ourselves, if only in our imagination, to the scenes of such events, i.e. to some of the world's most important deposits of Permian silicified woods: to Chemnitz (Germany), Nová Paka (Czech Republic), and Autun (France).
More than 290 million years ago in **Chemnitz** in the present-day German state of Saxony, an entire "forest" was preserved. Nowhere else but here did such a large floristic diversity of Permian plants occur *in situ*, in other words, at the location where they actually grew. For this reason, this locality provides especially favorably preconditions for the complex task of investigating well-preserved Permian floral elements. When the Zeisigwald volcano exploded NE of Chemnitz, its devastating eruption of glowing clouds killed all life in a small, tropical oasis. At this time, the descending ashes buried and conserved the remnants of a unique ecosystem, thus creating conditions that led to the formation of fossilized specimens. Today, these fossils provide us with a glimpse into an especially innovative moment in the history of plant

evolution. These Permian fossils ultimately attained international renown as the "petrified forest" of Chemnitz (Fig. 1). Petrified plants were and still are found at the base of the so-called "Zeisigwald tuff horizon" (Fig. 3). This stratum includes volcanogenic beds that were created in the wake of the aforementioned eruption. The tops of the fossilized trees point towards the west, in the direction the base surge was moving as it spread outwards from its epicenter during the first seconds after the volcano erupted. The tree trunks were instantly snapped and stripped of their branches and bark, but they fell and remained lying for subsequent petrification near where they had stood only moments before. Meanwhile, the pressure wave had carried their leaves and fructifications many kilometers further into the surrounding landscape. The genesis of the "petrified forest" of Chemnitz can be reconstructed based on comparison with the effects produced by the eruption of Mount St. Helens in May 1980. Thorough study of the volcanic deposits and the appearance of the vegetation after the eruption (Fig. 4) contributed a great deal to our understanding of the events that had occurred eons earlier in Chemnitz. Today, more than 20 years after the eruption of St. Helens, the peak's once densely forested hillsides show devastated surfaces which mutely testify to the events that transpired in 1980 and which lend support to the interpretation of fossil finds in the context of explosive volcanism.
Nearly complete tree trunks with roots or remains of branching are among the rarest specimens. Only the

3 Dadoxylon in tuff rock from Chemnitz
photo: Gerhardt, height 32 cm

4 Devastated vegetation after the eruption of Mount St. Helens in May 1980
photo 1994: Knorr

5 A construction excavation
with an exposed trunk
Chemnitz-Hilbersdorf, 1996

6 *Psaronius*-tree fern with a high percentage of blue-violet fluorite
collection of the Museum für Naturkunde Chemnitz, original size

most fortunate of the eager collectors have been lucky enough to discover such specimens at Chemnitz. The spectrum of preserved forms is diverse. There are both imprints of plants in volcanic tuffs and almost perfectly preserved specimens that show microscopically tiny structures of the plants' anatomy. Silicic acids leached out of the ashes and penetrated into the cells of the wood, where they solidified and preserved even the finest details of the cellular tissue. In some cases, the cells were preserved by bluish-violet fluorite (Fig. 6). The precise conditions and preconditions of this multiphase mineralization process remain to be elucidated by future generations. Only a beginning has been made in this venture thus far.

Nová Paka, located in an attractive landscape in the northern part of the Czech Republic, is often used as a synonym for a whole series of smaller deposits of silicified woods from the foothills of the Riesengebirge. Only those collectors who live in this region exactly know localities of the petrifications between Nová Paka, Stara Paka, Lisek, or Balka. The majority of the specimens are found in farmers' fields and, thus are rarely found in a clearly defined geological horizon. In its entirety, the so-called "Krkonoše Piedmont Basin" contains 1,800 to 2,500 m of non-marine alluvial deposits from rivers, swamps, and lakes, as well as intermediate horizons of volcanic origin. The layers in which silicified woods occur are interposed throughout the entire series of sediments and volcanites. We find the earliest of such fossils directly above the crystalline Variscan basement. Exposed in a quarry SW of Nová Paka, the gymnosperm woods are embedded in the coarse, reddish colored sandstones that comprise the Štikov arkoses of the Kumburk Formation. The horizontally lying specimens represent silicified driftwood and were deposited among erosive sediments from the underlying crystalline complex onto flat floodplains through which ran numerous small streams. The longest trunk ever found measures 8.2 m and is displayed in front of the museum in Nová Paka. This museum's comprehensive collection also includes specimens that suggest that other plant communities had also thrived nearby. Specimens of *Psaronius*, climbing ferns, and

7 The region around Nová Paka still offers the possibility to collect fossil wood fragments.

8 The museum in Nová Paka presents a noteworthy exhibition about the fossil treasures of the whole Nová Paka region.

Calamites trees, very similar to those found in Chemnitz, indicate the erstwhile existence of a moisture-loving plant community along the river banks of the Krkonoše Piedmont Basin (Figs. 9-11).
Sternberg, the cofounder of scientific palaeobotany, was already familiar with the first silicified woods specimens from Nová Paka. Research conducted by his young coworker, Corda, was also published in Sternberg's standard work, entitled "Versuch einer geognostisch-botanischen Darstellung der Flora der Vorwelt". Unfortunately, these early investigations did not continue during the 20th century, which explains why the interested collector today can search in vain for additional, more recent literature. Such texts simply do not exist. The focal point of palaeontological research in faraway Prague was based on other material, for example, fossils of the early Palaeozoic Barrandium or Tertiary basins of northern Bohemia. Fossilized woods of the Permocarboniferous still await investigation by future researchers. These specimens not only include silicified woods which found in the region's farmers' fields, but also from extremely fossil-rich cherts (Fig. 11). Both localities can be relied upon to yield interesting discoveries in the coming years.
Not far from Nová Paka, about 60 km further east, another deposit of silicified woods occurs within the Intra Sudetic Basin. This site is known in the literature as the "petrified forest" of Radowenz. The woods, the majority of which are of the gymnosperm type, are similar to specimens from the Kumburk Formation of Nová Paka or the Kyffhäuser in Central Germany (Rößler, in this volume). Derived from the Hexenstein arkose, which consists

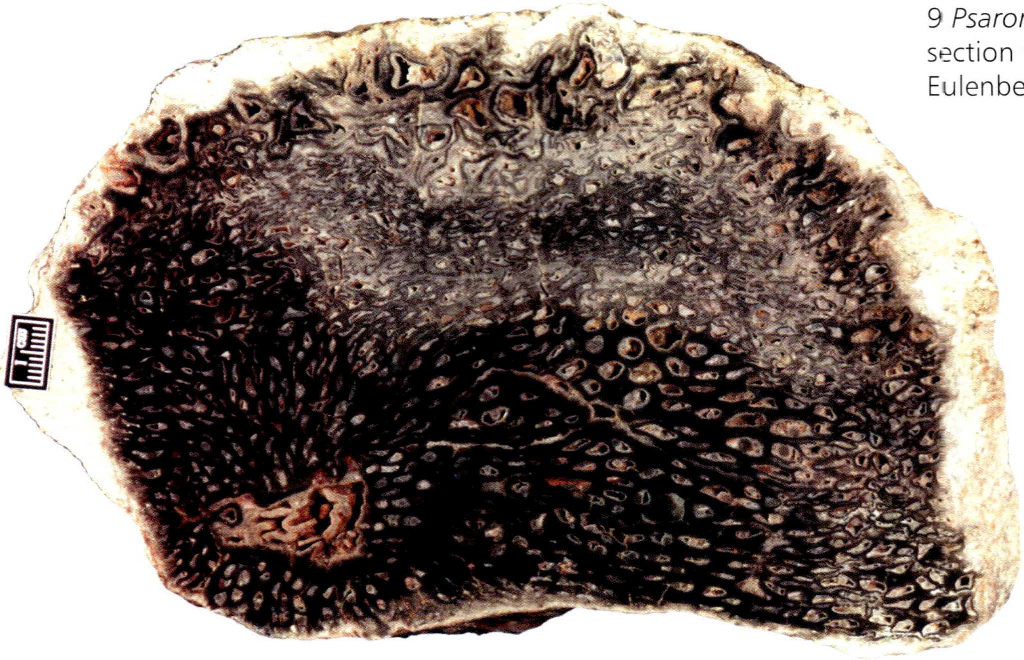

9 *Psaronius* sp., trunk in transverse section from Nová Paka
Eulenberger collection, Chemnitz

10 *Psaronius* sp., small trunk embedded in rhyolitic tuff from Nová Paka
Eulenberger collection, Chemnitz
original size

11 Fossiliferous chert from Nová Paka, collection and photo: Neuwald, Eggesin

of conglomerate and feldspar-rich sandstones from the Upper Carboniferous, these specimens first achieved fame through Göppert's 1857 publication. In addition to the trunks in the Breslau (Wroclaw) Botanical Garden, where Göppert once served as director, a present-day visitor to Radowenz (Radvanice) can also admire beautiful fossilized woods that have been included in the town's architecture, where they were used for decorative purposes.

The French city of **Autun** in the Massif Central primarily attracts visitors because of its 2,000-year history. The Roman city wall, the city gates of St. André and Arroux, and the Janus Temple (the largest Roman theater in Gaul), all testify to the formative influence which Roman culture exerted on the town's early history (Fig. 12). But also from a geological and palaeontological point of view, Autun is an important locality in the heart of France. Beginning in the

12 Roman-era Arroux city gate in Autun
photo: M. Barthel

13 *Arthropitys* sp., a small, anatomically preserved *Calamites* tree in cross section collection of the University Montpellier II, original size

14 Fossiliferous chert showing several axes of the climbing seed fern *Callistophyton* from Autun, collection of the University Montpellier II, original size

15 *Psaronius* sp. found around Autun collection of the Muséum d'Histoire Naturelle of Autun width 21 cm

mid 19th century, pieces of silicified wood have repeatedly been found in the fields around Autun, and these specimens have remained a focal point of scientific interest ever since (Figs. 13-17). After Brongniart, who described anatomically preserved sigillarias from this site in 1839, it was Renault who enriched our knowledge of fossil plants and fossiliferous cherts from Autun with numerous texts he wrote between 1868 and 1896. The objects of Renault's research were mostly based on discoveries made by a number of collectors and amateur researchers, including Roche, Lacatte, Rigollot, and Faivre. In later years, these collections provided basic material for studies by P. Bertrand and P. Corsin. Numerous investigations, especially ferns, can be credited to J. Galtier and his students. The most important specimens are housed in the Muséum National d'Histoire Naturelle in Paris; others are kept in the Muséum d'Histoire Naturelle in Autun; and still others are conserved at the Laboratoire de Botanique et Bioinformatique de l'architecture des plantes in Montpellier. In the late 1980's and early 1990's, construction near Autun of the TGV railroad line from Paris to Lyon uncovered the most recent exposures of silicified woods from this region.

The Lower Permian sequence of strata is known best from coal and uranium mining in the Autun Basin. Plant fossils are found in several horizons of bituminous shale (deposits from ancient lakes) into which tuffs are occasionally interspersed. In the years since Delafond published the first study of the area's stratigraphy in 1889, a total of four major discovery horizons have been documented in the Autun Basin. One particular difficulty has been that although the fossil remnants from the lowermost horizon belong to the Lower Carboniferous (Viséan), they can hardly be distinguished macroscopically from specimens found in the three overlying horizons. These include Igornay strata, whose age is between the Upper Carboniferous and the Lower Permian, as well as Lower Permian Muse and Millery strata. The majority of specimens occur in the Millery strata in the central part of the basin at Champs de la Justice, Borgis, and Espargeolles.

The qualitative floristic composition of the Lower Permian plant community at Autun is similar to the composition of plant communities found at Chemnitz and Nová Paka. Unfortunately, however, quantitative research has not yet been done at the two last-mentioned sites. According to a study conducted by Galtier and Phillips, the predominant plants at Autun were ferns and calamites, which together comprise approximately 84% of the specimens. Pteridosperms, including

16 *Colpoxylon aeduense*
A form which is very similar to *Medullosa leuckartii*, but without star-shaped vascular bundles in the pith, Autun, collection of the University of Montpellier II, 0.5 x

17 *Arthropitys* sp.
Calamites tree showing a thick woody stem from Autun, collection of the Muséum d'Histoire Naturelle, Autun

18 M. de la Comble, Jacques - former conservator at the Muséum d'Histoire Naturelle in Autun, photo: M. Barthel 1974

20 The "Rue Bernard Renault" in Autun, named after the famous French palaeobotanist

19 *Medullosa stellata*, Autun
collection of the University Montpellier II
original size

but not limited to such forms as *Medullosa* (Fig. 19), *Colpoxylon aeduense* (Fig. 16), Callistophyton (Fig. 14), account for about 8% of the fossil flora. A much less significant role seems to have been played by lycopsids (*Sigillaria brardii*) and conifers, whose fossils account for a mere 5% and 3%, respectively.

A quantitative study of this kind would be worthwhile for Chemnitz, but it would have to rely solely on newly collected materials because the majority of the older collections already represent "selections" undertaken with particular preferences and from various points of view.

MANFRED BARTHEL

Die Madensteine vom Windberg, Deutschland

The Maggot Stones From Windberg Ridge, Germany

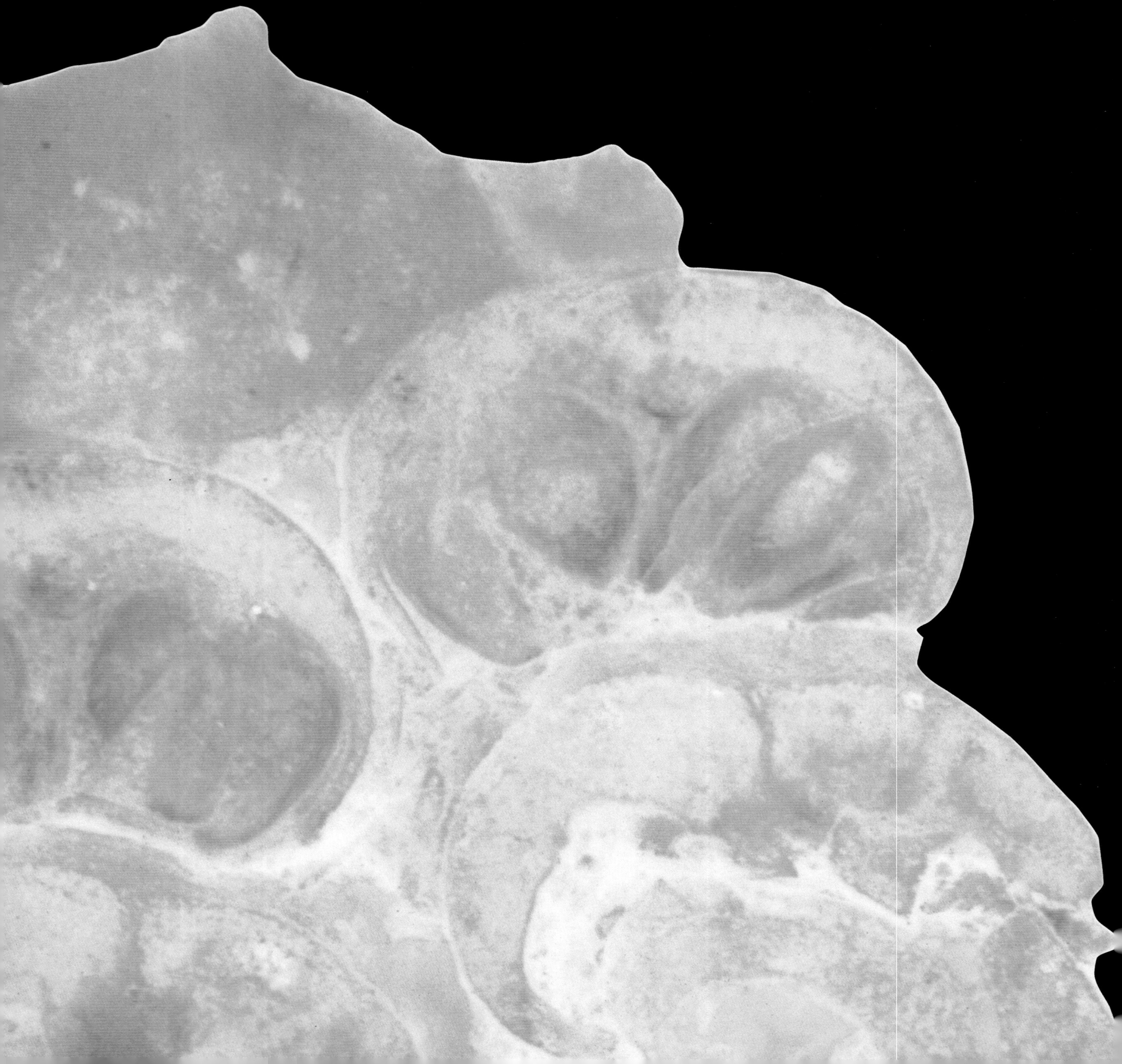

The Windberg, near Dresden in Upper Saxony, Germany, is the top of a woody ridge overlooking the industrial town of Freital (Fig. 1). The town is located in the densely populated valley of the Weisseritz, a small tributary of the Elbe River. Downstream, the valley narrows to a small canyon whose former wild scenery inspired the artists of the romantic period. The inhabitants of the valley have always regarded the Windberg as a true mountain and its fissures, old mountain buildings, monuments and legends seemed to me as real mystery when I was a young child growing up and going to school in the immediate vicinity of the Windberg. Later, as a student and as a palaeobotanist, I had the privilege to carry out several studies on the plant fossils from this area. I finally understood the scientific importance of the Windberg. The site has attracted the interest of geologists and palaeontologists for a long time. It was evident to them that the Windberg had been carved from a series of strata deposited in a small but deep Permian depression known as the Döhlen basin. The coal mining provided well-preserved plant compressions from the lowermost stratum, the Döhlen Formation. At the base of the ridge, the Niederhäslich Formation contains skeletons of Permian amphibians and reptiles which were found in 1881 while exploiting a limestone seam. On the flat ridge top, someone

1 The Windberg ridge near Dresden (Saxony), formerly a coal-mining district

discovered lumps of chert containing enigmatic inclusions from the uppermost stratum, the Bannewitz Formation. This is the setting of the story, which began about 240 years ago, went through several episodes of both amusing and scholarly fame, and is still going on today.

Not Maggots But Pinnules

The earthworks on the Windberg ridge, constructed during the Seven Year's War (1756-1763), had not yet been destroyed when Christian Gottlieb Pötzsch, custodian of the Electoral Natural History Collection in Dresden, inspected the place and found "among the thrown-up chert and petrified wood one agate-like specimen of red-brown and yellow-brown blending colors ... wherein some kind of petrified organic objects are lying in numbers ... whose former organic structure is clearly seen, being rather similar to a crooked sow-bug (Oniscus) ...." Later, in 1797, the Saxon officer and writer Friedrich Erich von Liebenroth found a piece of chert, apparently at the same site, with "more than a hundred insects of equal kind and size." The specimens were cut into a number of slabs by stone cutters in Dresden, who appropriately named them "maggot stones." Soon the slabs found their way by various means into several European collections.

The whereabouts of most of them are unknown today, but one (Fig. 2) with its original label is in the renowned Blumenbach collection at Göttingen University. The label indicates that as early as 1802 the botanist A.J.G. Carl Batsch in Jena, who was the Saxony-Weimar duke's garden supervisor and highly esteemed by Goethe, had recognized the true nature of the fossil. Sometime later, in 1835, another naturalist at Jena University gave a scientific description of the fossil which now is known as *Scolecopteris elegans* Zenker. Eduard Strasburger, the author of a renowned botanical textbook, thoroughly investigated the specimen in 1874. It was confirmed that the objects once compared to sow-bugs are actually the pinnules on

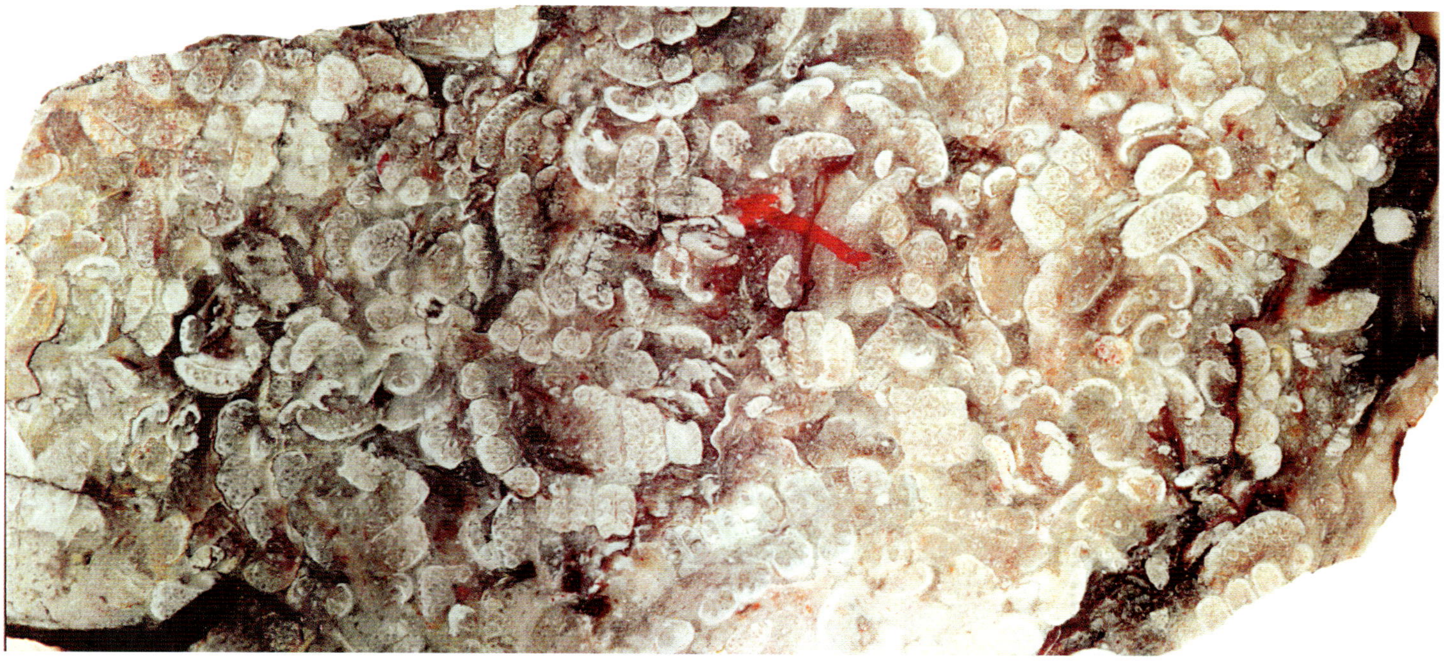

2 The oldest still available "maggot stone" find; collected by E. v. Liebenroth 1797 lectotype of *Scolecopteris elegans,* Blumenbach collection at the University of Göttingen, 3 x

3 The largest and most excellent specimen of maggot stone for 240 years was found last year by Dr. W. Schwarz (Dresden), polished section of nearly 190 x 190 mm

a large frond of a *Psaronius* tree fern. In the nomenclatural sense, it is the type species of *Scolecopteris* (literally meaning "maggot fern"), a common and widely distributed genus in Upper Carboniferous and Lower Permian floras.

Once in a Century

More than a hundred years after the first finds, around 1890, a few more "maggot stones" were ploughed up in the fields of the area and were collected by geologists from the Saxonian Geological Survey. In one of the several coal mines beneath the ridge, one small chert seam was discovered, consisting of a different chert but with the same maggot fern pinnules. Again, about a hundred years would pass without further discovery. In scientific literature, the Windberg site was referred to only as the type locality and its find as the type specimen.

After 240 Years - The Best Comes Last

Beginning in 1985, a series of lucky circumstances have occurred, starting with mapping work by the mining geologist Wolfgang Reichel of Dresden that led to an unexpected revival of collecting "maggot stones". Wondering where the original locality was, he considered potentially suitable sites for the old military encampment and its earthworks. He then came to the conclusion that the east end of the ridge was such a

3a Detail 125 x 90 mm
photo: Schwarz

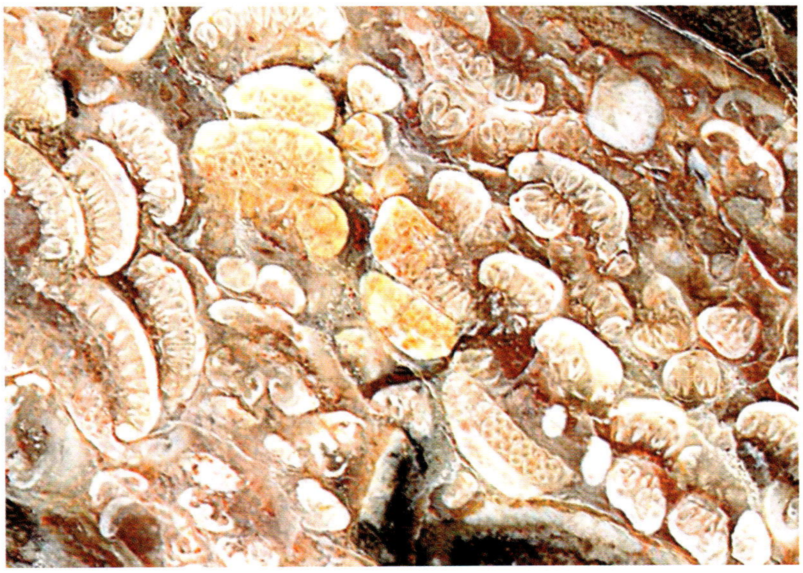

3b Detail with isolated fertile *Scolecopteris elegans* pinnules in different sections
22 x 16 mm

3c Detail with a sterile ultimate pinna
21 x 14 mm

4 *Scolecopteris* synangia mineralized by chalcedony (white), hematite (red) and goethite (yellow), 23 x 15 mm

place. It was there that his coworker, the miner Gert Müller of Dresden, found several chert pieces containing poorly preserved *Psaronius* stems and one beautiful "maggot stone".
This new find, in addition to the earlier ones, encouraged Hans-Jürgen Weiss of Rabenau to scan the ploughed fields around the Windberg and surrounding areas. He and his wife found several new sites in Pleistocene fluviatile deposits with sparsely distributed "maggot stone" cherts. Housing construction on the Windberg ridge since 1994 brought more cherts to the surface.
The Weiss collection is now the most important collection of these silicified plant fossils from the Bannewitz Formation.
The "maggot stone" story reached an unexpected highpoint in the summer of 2001. The mineral collector Wolfgang Schwarz of Dresden, watching a power shovel working at the famous site, noticed a layer of big, closely packed stones slightly below ground level. They were obviously put there on purpose, possibly to make a strong platform for a cannon at the old military encampment. From this layer of no more than six square meters, he dug out the most beautiful "maggot stone" ever found (Fig. 3).

Where Do They Come From ?

It must be mentioned that cherts of various types are present in every one of the three main formations of the basin, but the "maggot stone" cherts are restricted to the Bannewitz Formation. They are usually found as more or less rounded fragments of chert layers or lenses whose original position and extension is not exactly known, with the one exception of the red chert found in a mine as mentioned above. There are also two sites with sharp-edged fragments only, indicating that the original layer of the red chert must be nearby.
A rhyolitic tuff deposit 20 to 40 meters below these localized red chert layers may have contributed to their formation. The chert pebbles found on the surface, including the classical "maggot stones", are part of

a light-colored fluvial deposit consisting of clay, sand, gravel, and boulders in various proportions, extending over an area as small as three hectares. According to the geologist Karl-Armin Tröger of Freiberg, this is a locally preserved remnant of the Upper Cretaceous Niederschöna Formation, deposited before the transgression of the Cenomanian Sea, whose sandstone deposits are visible nearby.
Our "maggot stone" cherts are Lower Permian, lithostratigraphically Upper Rotliegend. As long as there is no more precise dating, we assume an age of about 290-280 million years for these deposits. Because they have not yet been subjected to petrographical and geochemical tests, no details are known about their origin. However, it seems certain that the chalcedony was formed by silica (SiO_2) deposition in a lacustrine environment, as indicated by sediment structures and the inclusions of tiny shrimp-like fossils. Conversely, *Psaronius* roots, sphenophylls, and *Calamites* indicated a wetland habitat. The various types of chert containing *Scolecopteris* can be assumed to have originated under different conditions. Volcanic and post-volcanic activity associated with hot springs is the probable source of the silica solution, as known from other European sites with Carboniferous and Permian cherts and silicified wood. Diverse ferric minerals, mainly hematite and goethite, impart color to the chert (Fig. 4). Small chalcopyrite crystals with shining faces are occasionally found within the transparent chalcedony. Dark layers in the chalcedony may contain residual organic matter.
The typical "maggot stone" aspect of the chert is due to densely packed pinnules, pinnae of ultimate and penultimate order. There is very little difference between the sterile and fertile pinnules. They are certainly

5 *Psaronius infarctus* Unger
probably the trunk of the "maggot-fern"; a central part of the stem with vascular bundles and sclerenchyma bands
48 x 41 mm

mature stages of trophosporophylls. Because the chert contains parts of the fronds, fructifications (*Scolecopteris*), petioles (*Stipitopteris*), trunks (*Psaronius*; Fig. 5), and roots, it provides a rare opportunity to investigate a fossil plant as a whole. Thus far, investigations have been restricted to visual observation of cut and polished sections with the reflected-light stereomicroscope. Because of the poor optical contrast, the customary thin-section technique is of little use. The surface-peel technique has not been tried because of the apparent homogeneity of the chalcedony. It might, however, prove useful with the newly collected, well-preserved, but less dense silicified specimens. Tissues having thick-walled cells, such as those on the outer surface of the sporangia, the vascular tissue, sclerenchyma strands, and cortex, can be frequently

6 Ultimate pinna of *Scolecopteris elegans* with several xeromorphic features; attached pinnules bent downward and faced with their lower sides, 6 x 9 mm

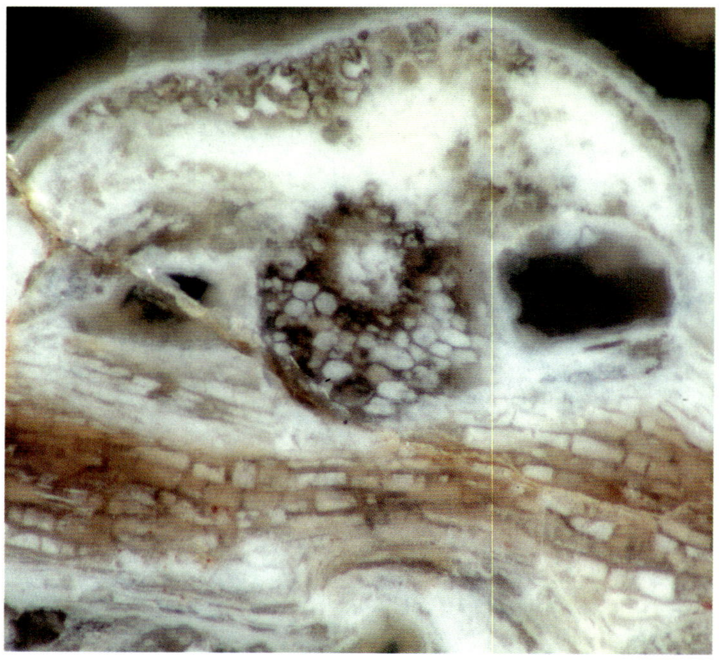

6a Transverse section of a sterile pinnule in detail; note the extremely thick lamina and midrib; below it are the cortical cells (cortex) of attached pinna axis in longitudinal section, 2 x 2 mm

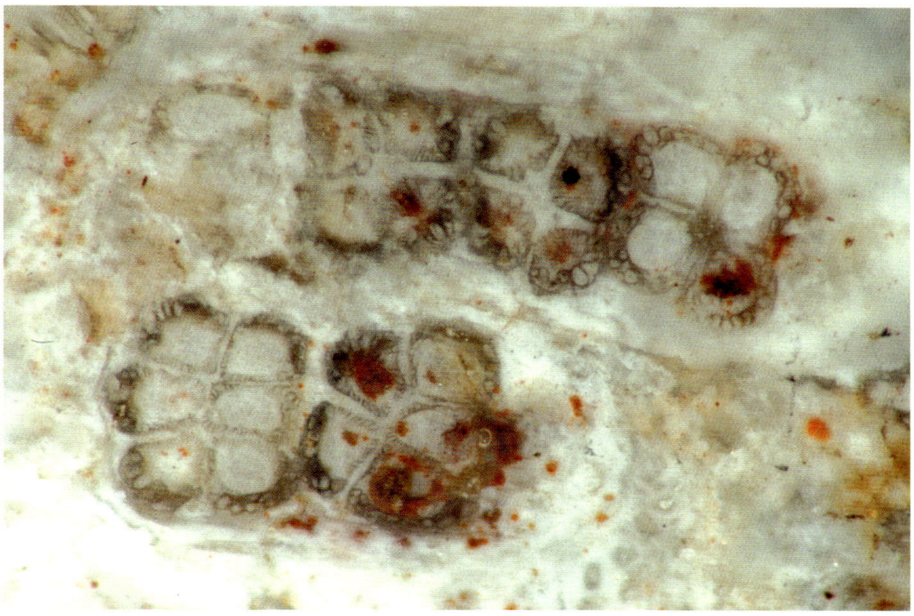

7 Fertile pinnule of *Scolecopteris elegans* sectioned in a paradermal plane; varying number and symmetry of sporangia in the synangia; sporangial walls outwards prominent, inwards thin, 3 x 2 mm

observed under reflected light.
With the larger amounts of chert available, specimens display well-preserved soft tissue, such as palisade parenchyma within the pinnules and the air-filled parenchyma of *Psaronius* aerial roots. Even the sub-cellular scalariform structure of tracheid walls is occasionally visible.

Ferns in Bright Sunshine - Survival Strategies

Pinnules showing several xeromorphic features can be seen on the polished surfaces and even on the natural surface of the "maggot stones".
A layered mesophyll makes the pinnules thicker than other fern pinnules of comparable size, and sometimes their midvein is extremely thick and bulging (Figs. 6, 6a).
The pinna axis, too, is thicker than expected. The margin of the fertile pinnules is usually curved downward to protect the synangia (Fig. 3a), often with tapering fringes for additional coverage, or the sporangia protect themselves by forming thick capsule walls. Occasionally, epidermal hairs are attached to the midvein below the pinnule (Fig. 3c).
As is known from modern xerophytes growing in salt marshes, such features are not necessarily related to dry soil conditions. In the case of our maggot ferns, these features may be due to an exposed habitat with high insolation and dry winds, combined with an inefficient vascular system unable to cope with the situation. One should bear in mind that the climate during the Upper Rotliegend was much drier than that of the rain forests and the misty mountain forests which are preferred by the recent tree ferns. Also, there seems to have been no shelter available from larger trees such as *Cordaites* at the maggot-fern sites.

A Variable Species

As with every *Scolecopteris* species, our maggot fern has synangia consisting of 4 or 5 (or sometimes as many as 7) sporangia attached to a common pedicel. The synangium tapers to an acute tip. The number of sporangia varies even on one pinnule (Fig. 7). Their arrangement may vary from radially symmetrical to poorly symmetrical, so that symmetry is not a characteristic feature, contrary to claims in the literature. Other features may also vary within wide bounds, such as the shape and size of sporangia and the thickness of pinnules. Such variability may be due to ontogenetic and/or environmental

8 An ultimate pinna of *Scolecopteris elegans* with well-preserved surface of the pecopterid foliage
11 x 8 mm

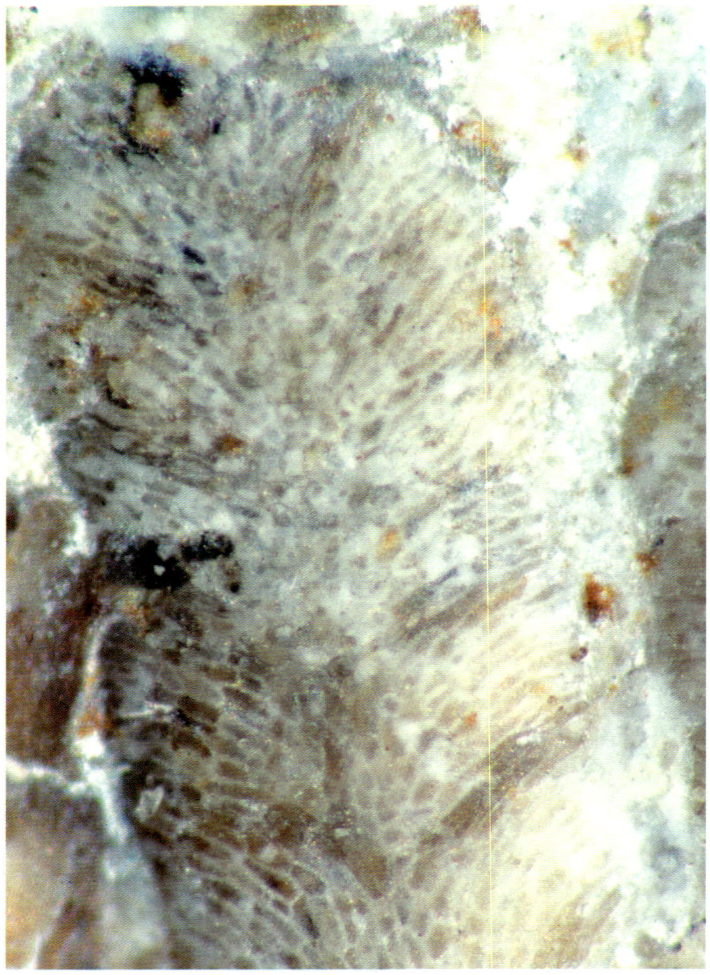

8a Distinct cells of upper epidermis in detail
3 x 2 mm

causes. With the possible exception of a few problematic specimens, the tree-fern foliage of the "maggot stone" cherts belongs to *Scolecopteris elegans*.

A Holistic Approach

It is difficult to relate species of fern pinnules to other plant remains fossilized in different forms of preservation, such as permineralized to compressional states. Thus the question arises of how to relate our permineralized *Scolecopteris elegans* to a certain species of the compressional form-genus *Pecopteris*. Fortunately, there are some "maggot stones" with pinnae lying flat on the surface of the chert layer with their venation clearly visible. The lateral veins are non-branched and nearly straight, which is typical of *Pecopteris arborescens* Brongniart (Figs. 8, 8a). This latter species has also been found in the uppermost stratum of the Döhlen basin, embedded in the siltstone above the red chert layer. From specimens from other localities, we know that *P. arborescens* has large tri-pinnate fronds, which probably applies to our *Scolecopteris* as well. Because of the much smaller size of the chert specimens, the largest fragments found are pinnae of penultimate order.
It has long been known that *Scolecopteris* is one of the fertile foliage types associated with *Psaronius* tree trunks. Evidence for this has been provided by specimens from other coal basins which have yielded trunks with petioles (frond stalks) still attached. However, the trunks were poorly preserved, thus

making it impossible to establish relationships between the trunk species and foliage species. This task is made more difficult by the fact that there are usually several species present at the same locality.
The situation is less ambiguous in the Bannewitz Formation of the Döhlen basin, where only one species of *Scolecopteris* appears to be present. A number of chert specimens show *Scolecopteris* pinnules directly beside *Psaronius* stems or roots, hence there is little doubt that they belong together. Unfortunately, most of the stems are flattened so that the details of their anatomical structure are difficult to see. However, a few of the less flattened stems show numerous irregularly arranged vascular and sclerenchyma strands in their centers surrounded by a sclerenchyma sheath.These features are typical for *Psaronius infarctus* Unger (Fig. 5). Therefore we can assume that the stem of our maggot fern may be included in this well-known and widely distributed tree trunk species.
Cherts can preserve interesting biotic relationships and associations. Younger Palaeozoic cherts, preserving fossil plant remains, are known from several European sites, such as Sardinia, Italy; St. Étienne and Autun, France; Chemnitz-Altendorf, Germany; and Nová Paka, Czech Republic, among others. There is a maggot stone from the Windberg site with tiny burrows gnawed into the cortex of the petioles, with coprolites inside, probably produced by mites or other arthropods (Fig. 9). Similar coprolites

9 The cortex of a *Scolecopteris* frond axis in transverse section partly destroyed and filled with coprolites produced by oribatid mites or other arthropods, 3 x 2 mm

are abundant in the adventitious roots of the large psaronia from the famous Chemnitz site (Rößler, in this volume).
In addition to the dominating maggot fern, a few other species are sparsely distributed in the cherts of the Windberg site. These consist of slender axes of the herbaceous swamp plant *Sphenophyllum* with partially preserved central steles and cortices (Fig. 10), and trunks and rootlets of the calamite *Arthropitys* (Fig. 11). Petrified wood pebbles are abundant at the site, mostly of some *Dadoxylon* type that were flattened before petrifaction. *Cordaites* leaves and "*Walchia*" twigs have not been found in the Windberg cherts, but occur in other nearby areas.

The Power of Large Numbers

Now that hundreds of well-preserved specimens have been found and the "maggot stones" from Windberg are no longer historical rarities and taxonomic reference samples, one wonders whether they have lost their status as something special.
The large number of samples of the same species from the same site provides evidence for natural variability. The "maggot stones" reveal rich structural information due to their three-dimensional preservation. The favorable conditions found at Windberg are rarely encountered in the collecting and study of fossil plants (Rothwell, in this volume). Because of their unique preservation, the "maggot stones" from the

10 Small stem of *Sphenophyllum* with its central triarch stele and surrounding cortex, 5 x 3 mm

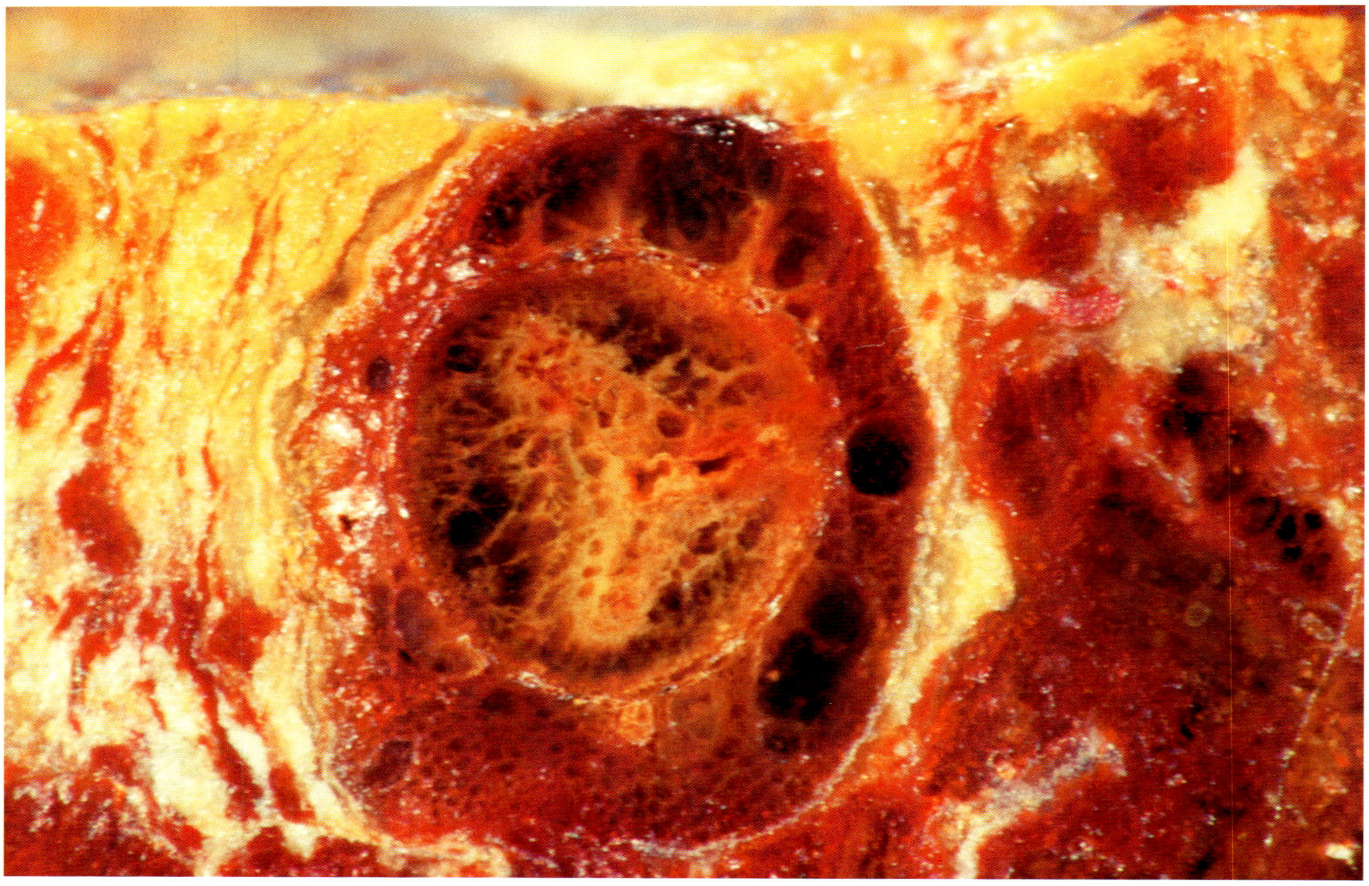

11 Transverse section of the calamite tree *Arthropitys*
width nearly 55 mm

12a, b An Austrian "Kreuzer" coin minted in 1762
Its discovery in the present maggot stone site in Freital-Burgk proves that this locality is identical with the historical site in the Seven Years War (1756-1763).

Windberg definitely remain a palaeobotanical treasure.

There's No End…

Most of the cherts collected from the Windberg site are still awaiting cutting, polishing, and evaluation. Even with the reduced number of new collections expected in the future as development on the Windberg site comes to an end, we are confident that the chert pieces which have already been recovered will reveal additional details and thus contribute to a better understanding of ancient ecosystems.

Recently, a construction worker found an Austrian coin on the site with Empress Maria Theresia on one side and "1762" on the other (Fig. 12a, 12b). The coin was possibly lost by a soldier during the Seven Years War while using the "maggot stones" for a fortification.

ULRICH DERNBACH
ROBERT NOLL
RONNY RÖßLER

Neues von Araguaina, Brasilien

News From Araguaina, Brazil

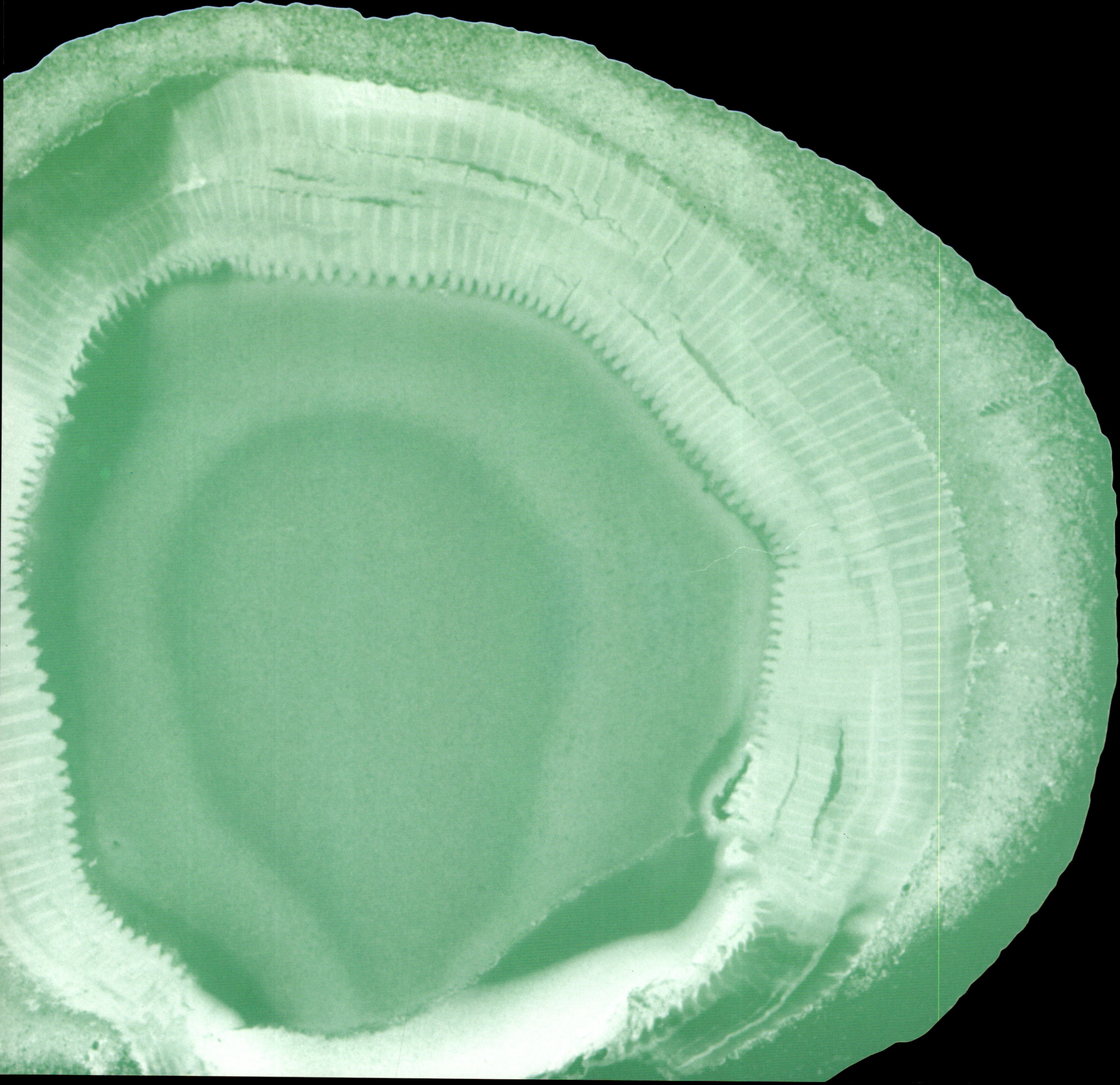

We first visited Araguaina in 1994. In the book "Petrified Forests" one can read the story of how we found the first fossilized *Calamites* tree at that locality. Prior to this point in time, one knew only about the tree ferns, and especially about the ones which were available on the local market under the names of "Samambaia" or "*Psaronius brasiliensis*". This *Tietea singularis* (which, incidentally, was first discovered near Tieté in southern Brazil in 1907), has remained the most commonly found fossil from the Permian of Brazil. In 1995 we found *P. brasiliensis* and other psaronias (e.g. *P. sinuosus*). These finds were augmented in 1998 by spectacular discoveries in Araguaina of fossil *Calamites* plants, including one particularly remarkable calamite specimen measuring more than two meters in length and nearly 20 centimeters in diameter.

A Critical Situation

While we were unearthing this woddy calamite trunk, my friend and constant companion, Decio Freitas, was stung by a scorpion. Rather large numbers of these arachnids "keep watch" over the fossil treasures of the moist savanna. Caution is necessary, but then again, who wants to wear gloves on days when the mercury climbs above 35°C (95°F)? Within minutes, the girth of Decio's hand had swollen to twice its normal size. A very critical situation had arisen, especially since we didn't have any medication that could counteract the scorpion's poison. To relieve the stress on Decio's circulatory system, we sat him down in the shade and did our best to cool his hand with ice. Not until several hours later, when we had begun our return journey, did the swelling begin to subside.

1 Decio Freitas at the petified forest locality
He was stung by a scorpion.

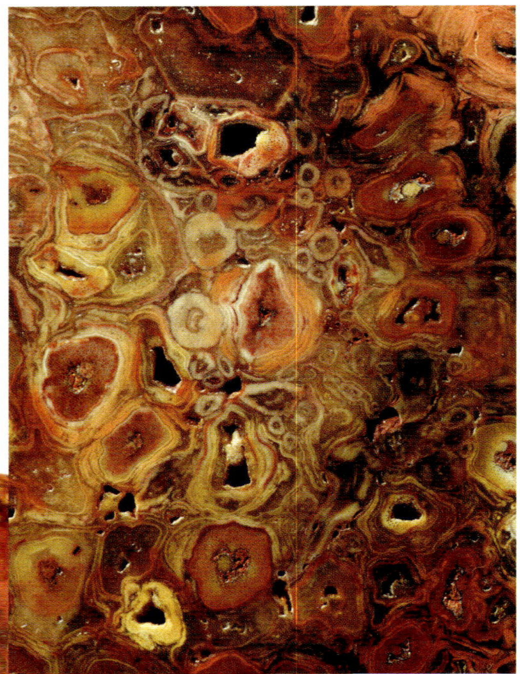

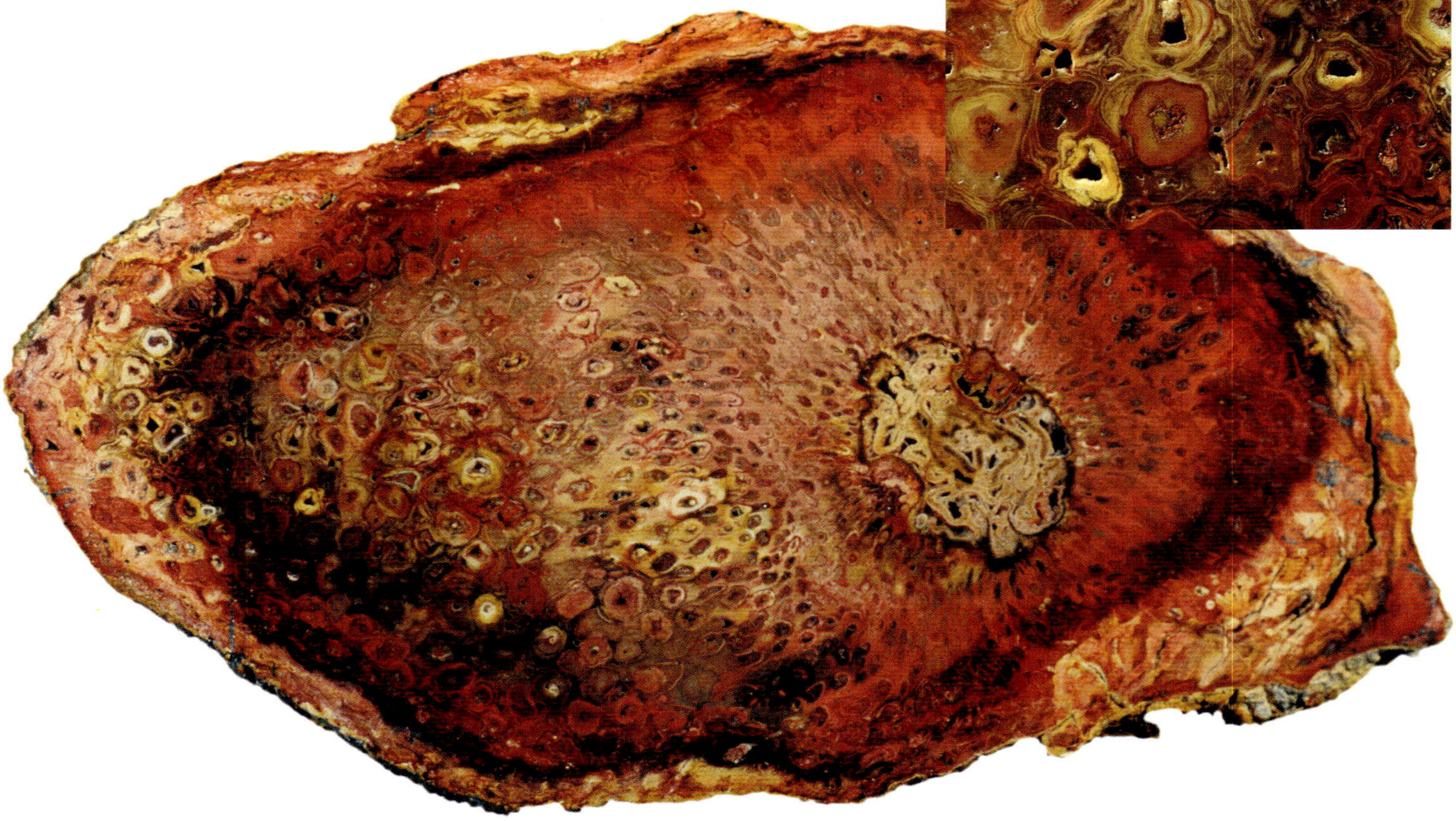

2, 2a *Psaronius brasiliensis* showing several small petioles of a new epiphytic botryopterid fern growing inside the outer root mantle of the tree fern
Dernbach collection, 31 cm diameter

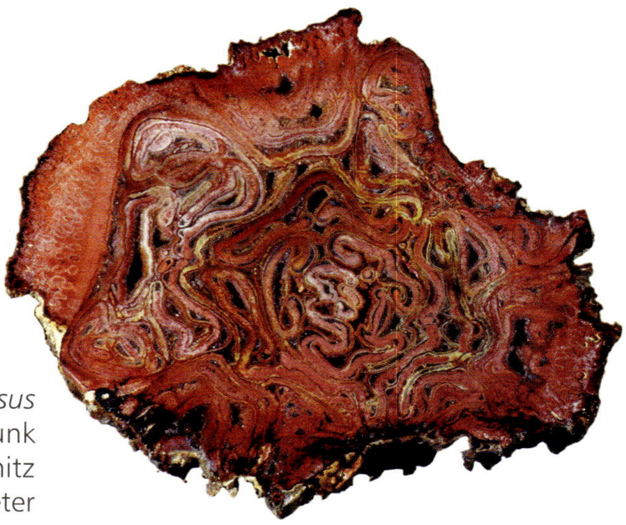

3 *Psaronius sinuosus*
Transverse section from the upper part of the trunk
collection of the Museum für Naturkunde Chemnitz
9 cm diameter

4 Landscape between Araguaina and Filadelfia showing the sandstone cliffs of the Pedra de Fogo Formation

Spectacular Finds

We had already discovered the first remains of fossil fern foliage in 1999. But thus far, no one has unearthed specimens with attached fronds. The only finds of this type to date are fossils of individual pinnae embedded in sandstone. Presently, based on morphological criteria or on examination of the attached sporangia, we can differentiate three different forms, but we're still unable to assign them with any degree of certainty to their erstwhile stems. Chert lenses have yielded additional isolated fragments of ancient plants, including the axes of sphenophylls, which we recognized as familiar "old friends" from the Northern Hemisphere.

In 2000, I was accompanied on my trip to Araguaina by hobby palaeontologist R. Noll and R. Rößler, who directs the Museum of Natural History in Chemnitz. Both men are well acquainted with Permian floras, so it was only reasonable to invite them to accompany me to Araguaina. During our nearly weeklong expedition, we found several excellent specimens of the extremely rare tree fern, *Grammatopteris*. The largest of these specimens measure 20 cm in diameter. Prior to this discovery, *Grammatopteris* had been found only twice: once in Autun, France, towards the end of the 19th century, and the second time in Chemnitz, Germany, in 1915. Even more extraordinary is a specimen which we were fortunate enough to find on the final day of our expedition. I had discovered a thus-far unknown tree fern which combines ancient and modern anatomical characteristics in a single plant. These discoveries are currently undergoing scientific study at Chemnitz.

5 The expedition team and their local guides from Brazil

6 *Arthropitys* sp.
Woody *Calamites* tree in transverse section showing secondary xylem and rays
Noll collection, width 2.4 mm

7 *Arthropitys* sp., detail showing the delicate tissue of the pith (upper part) and the xylem groups (lower part)
Noll collection, width 3.6 mm

8 Calamites trunk showing the segmented pith
Dernbach collection, height 30 cm

The Petrified Forest of Araguaina, Considered From a Geological Point of View

To set out in search of fossils in Brazil is by no means as simple as it may sound. More than half of the country's area is occupied by elevated shields of very old rock complexes entirely devoid of fossils. It is these shields which, together with amphibolites, granulites, gneisses, and granitoids, form the firm basement of the continent. They date from Precambrian eras when life on Earth had not yet ascended beyond the stage of unicellular microorganisms.

The Maranhão Basin in NE Brazil, in which our expedition's territory belongs, stretches across more than 600,000 square kilometers (230,000 square miles). Alternately marine and continental sediment series from the Silurian to the Cretaceous have been deposited here. The chief collecting horizon in the petrified forest is the Pedra de Fogo, the so-called "fire cliff" formation. Mapping has shown that this complex of strata ranges between 25 and 200 m in thickness and covers an enormous area spanning parts of the Brazilian states of Pará, Tocantins, Maranhão, and Piauí. Important discovery sites are located in the northeastern part of the State of Tocantins, SE of Araguaina, in the area between Bielandia and Filadelfia, as well as to the east of Colinas.

A Lower Permian age is assigned to the Pedra de Fogo Formation. This dating is supported by palaeontological evidence, e.g. discoveries of labyrinthodont amphibians and a diverse Lower Permian microflora. The environment in which these strata were deposited seems to have been a river system with expansive lakes and extensive lagoon regions along its periphery. Occasional advances by the ocean quickly flooded lower lying regions, during which it deposited carbonates and anhydrites. We can also assume that the climatic conditions must have been correspondingly variable, with wide oscillations from humid to semiarid.

How Was the Petrified Forest in Northeastern Brazil Created ?

Erstwhile vegetation could perhaps have lined the banks of watercourses throughout an extensive ancient river system. Here there would have been anastomosing shallow streams with pebbly bars and sandy walls, meandering within an extensive plain dominated by countless lakes and ponds. Moisture-loving plants, especially tree ferns and various types of horsetail-like plants, probably

10 *Grammatopteris freitasii* Rößler & Galtier juvenile trunk, paratype, collection of the Museum für Naturkunde Chemnitz original size

9 *Grammatopteris freitasii* in transverse section, Dernbach collection, 16 cm diameter

11 The newly discovered small tree fern *Dernbachia brasiliensis* Rößler & Galtier in transverse section; holotype; collection of the Museum für Naturkunde Chemnitz original size

11a Petiole in transverse section 3 x

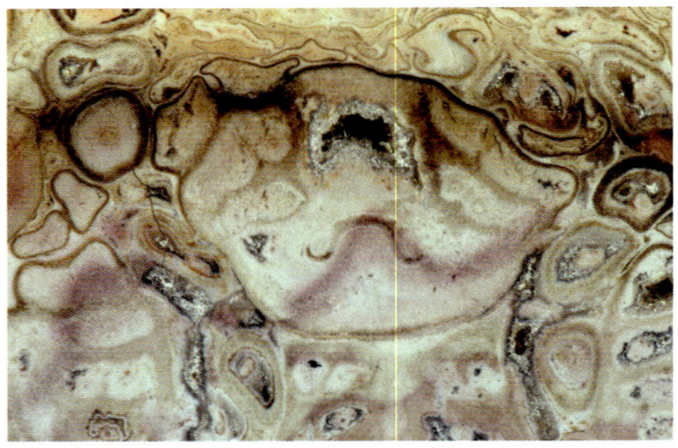

inhabited the more or less broad, lushly vegetated shores. In the wake of torrential rainfalls, the rivers overflowed their banks and carried away parts of the plants in their floodwaters. Vegetative materials were no doubt deposited within the coarse clastics on the bottoms of the shallow streams. Longer fossil stems show an unmistakable tendency towards deposition in a NE-SW direction. This can be interpreted as an indication that the general transport direction in this Permian river system may well have been northeasterly. The silicified woods were not deposited in situ, i.e. they must have first been swept away from the sites where they originally grew and then deposited at greater or lesser distances from those locations. The movement, however, could not have been inordinately lengthy, because rather than finding only isolated trunks, we also find remains of branches, foliage, and roots. All of these provide evidence against the hypothesis of a longer transport path and subsequent dispersal of plant parts during the journey. The delicate pinnae of the fronds of tree ferns are frequently found preserved in the thin, sandy layer along the peripheries of the trunks. Erosive products rich in silicic acids later penetrated the trunks and mineralized the cellular tissue as well as the pore spaces in the embedding sandstones.

The Composition of the Petrified Forest

Closer examination shows that the flora was by no means as uniform as claimed in the literature. This appears to be confirmed by the monotony of the specimens available at mineral fairs and markets in Brazil. In addition to *Tietea*, *Psaronius* also occurs frequently. Unfortunately, the basal sections of the trunks of these plants, which bore dense mantles of adventitious roots, have rarely been collected to date. The material offered for sale at various places creates the misleading impression that the tree ferns of Gondwana must have been smaller and less anatomically complex than their relatives in Euramerica. During the course of our expedition, we discovered convincing evidence that this was not necessary the case. We now have reason to believe that fern trunks measuring as much as one meter across at their base were by no means uncommon. As far as branching is concerned, there is an obvious dominance of the so-called "four rowed" type, which includes both *T. singularis* and *P. brasiliensis*. The first discoveries of petioles (*Stipitopteris*) and permineralized pinnae now complement the tree fern trunks.

Several new tree ferns were discovered for the first time in 2000. These include a new species (*G. freitasii*) of the very rare *Grammatopteris* genus which had thus far included only two species (*G. rigollotii* und *G. baldaufii*) and had only been known from specimens found in Europe. The growth form of these plants is

12 Outer surface and longitudinal section of a coniferous stem
Note the typical sclerenchyma nests in the pith
Noll collection, height 8.2 cm

13 Conifer stem showing well preserved tissue in the central pith
collection of the Museum für Naturkunde Chemnitz, width 3.6 cm

14 Gymnospermous stem in transverse section completely surrounded by silicified sandstone
Noll collection, original size

comparable with that of the psaronias, and they were supported at their bases by dense mantles of adventitious roots. These mantles taper and become smaller as they rise upwards, and are entirely lacking at the upper portion of the trunk, where one sees the petiole scars which closely surround the trunk per se. The adventitious roots are relatively small and diarch in construction, which makes them comparable with those of many other fossil and recent ferns. The discoveries made it possible to study different growth phases ranging from small examples measuring just 4 cm in diameter to larger specimens whose basal cross-sections measure nearly 40 cm across. The trunk consists of a simple exarch protostele and a complex, structured cortex in which various characteristics (e.g. reinforcing sclerotic nests around the leaf traces or secretory cells) could be identified. The petioles with their typically columnar bar-shaped bundle make it possible to determine that the pinna axes are arranged in two rows, but the foliage remains to be discovered by future researchers. All in all, it was of major importance to discover that *Grammatopteris* had indeed been present in the Permian of South America. *Grammatopteris* may have been a possible link in the early evolution of the Osmundaceae.

Another discovery has brought to light an entirely new tree fern. Its unique and still unknown mosaic of characters unites a large actinostelic type of stem with a modern dorsiventral type of petioles. The steles are similar to those of the genus *Asterochlaena*, but the petioles are more like extant, contemporary forms. The fronds grew in a variable number of orthostichies, straight rows along the stem, and the number of fronds corresponds to the number of xylem arms of the stele. The morphology of the petiolar xylem of these fronds

15 Well preserved tree fern (psaronia) pinna
Dernbach collection, width 80 mm

16 Diverse plant remains three-dimensionally preserved in silicified sandstone
Noll collection, width 36 mm

17 Tree fern pinna with sporangia densely filled with spores, Noll collection, width 6 mm

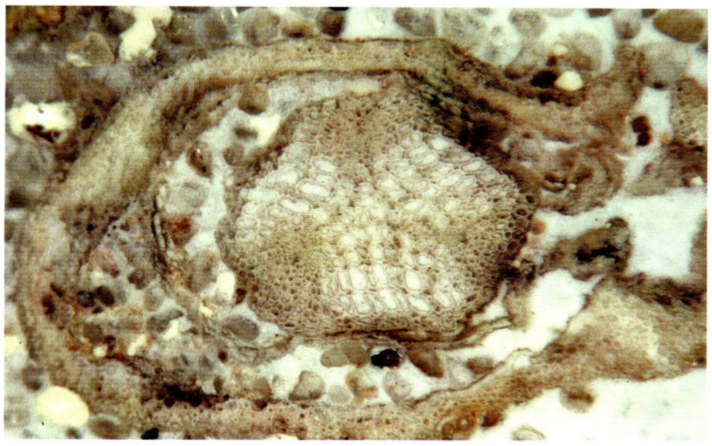

18 *Spenophyllum* axis in transverse section showing the central triarch stele and cortex remains
Noll collection, width 6 mm

19 Gymnospermous root in transverse section
Dernbach collection, width 8,4 mm

undergoes important changes in their individual development. Leaf traces depart as bipolar strands, enlarge tangentially, and finally become Pi-shaped. Numerous adventitious roots occur between the petiole bases. Thus far, few individual specimens show this fern in different states of preservation and, based on its differences from other known forms, justifies the introduction of a new genus named *Dernbachia* (on the cover of this volume). The foliage and fructifications of this fern remain unknown and hopefully will be discovered in the future. The collected material was recently studied and published.

At first glance, the woody "horsetail trees" (*Calamites*) appear to be identical to the European specimens of *Arthropitys*. Our expedition discovered such specimens up to 3 m long with trunks up to 20 cm thick. Among the *Calamites* specimens, it is possible to distinguish several types by taking into consideration their anatomical features and branching patterns. But a great deal of detailed scientific work must first be done before their taxonomy can be determined.

Dadoxyla, the gigantic trunks of gymnosperm trees, are not only as common in the Permian of Araguaina as they are in the Permian of Europe, they have also been overlooked because most studies have centered on the exotic ferns and calamites. Even from a scientific point of view, very little research has been conducted thus far. We therefore decided to search for gymnosperm trunks and, of course, for anatomical features that would support their identification. Such features include, for example, the structure of the pith and primary wood, the fine structures of the secondary wood, the leaf traces, and the mode of branching. In the field, we found gigantic trunks, as well as countless fragments of root wood and broken pieces that will remain undescribed for the present. Whereas one single specimen clearly showed sclerotic nests in its pith, and could thus be unmistakably identified a piece of conifer wood, we found in it the doubled leaf traces that are often associated with cordaites. The mode of branching suggests a high degree of variability, which tells us to avoid hurried comparisons and overly hasty misidentifications. Not only did we frequently find "whorl-like" branching of a type like "walchias", we often observed boughs with the leaf traces arranged along the trunk in loose, diagonal rows – a system of branching that has never been observed before among early conifers, nor did it ever appear in later forms of coniferous trees.

Koniferen aus den „Uplands" - Permische Kieselhölzer aus der Mitte Deutschlands

Conifers From the "Uplands" - Petrified Wood From Central Germany

Petrified wood frequently offers an attractive combination of a plant fossil with precious agate.
Such eye-catching petrifactions have been collected for a long time. Though not rare in some parts of Germany, they are especially abundant in areas characterized by the outcropping sediments of the Lower Permian Rotliegend facies.
Our interest in petrified wood of the *Dadoxylon* type is based on the collection of well-preserved specimens from the Lower Permian Donnersberg Formation of the Palatinate (Pfalz) west of Heidelberg and the Schöneck Formation of the Wetterau north of Frankfurt am Main. These consist primarily of the wood of early conifers from the so-called hinterland.
If not otherwise indicated, the illustrated specimens from the Palatinate are kept in the private collection of the first author; those from the Wetterau are housed in the collections of the Forschungsinstitut Senckenberg, Frankfurt am Main.

History and Circumstances of Discovery

The first specimens of petrified wood from the Rotliegend facies of the Wetterau were described from the Naumburg near Kaichen by Göppert in 1864/65 in his classical monograph on fossil conifers. The localities of Bad Vilbel and Schöneck-Kilianstädten also belong to the Wetterau, a lowland area northeast of Frankfurt am Main. The history of investigations of wood petrifactions from the Palatinate began with Schuster in the early 20th century. Later, Frentzen included material from the same area in his

2 Transverse section of a conifer stem
with conspicuous pith surrounded by
wedges of primary xylem
Kilianstädten, 140 x 92 mm

1 Petrified wood with agate and amethyst
260 x 200 mm

3 Landscape near Winnweiler

monographic treatment of *Dadoxylon*, and newly collected material of anatomically preserved conifers from the Donnersberg Formation was recently described by the authors.
But these will surely not be the last discoveries! It is still possible to collect interesting specimens practically on one's doorstep. Freshly ploughed fields and construction sites are the most promising places to search.
As in the past, most of the private collectors' interest is still focused on recovering large and spectacular specimens of stems. Less obvious fragments of twigs, occasionally showing anatomically important internal structures, are rarely collected and thus underrepresented in the collections. But it is precisely these smaller items which may be of the greatest scientific value.
Many species of *Dadoxylon* have regrettably been described on the basis of insufficient and sometimes poorly preserved material. But proper understanding and description of a fossil species and its variability depends upon collecting, scrutinizing and puzzling over as many specimens as possible. Another advantage of thoroughly collecting at individual localities and clearly distinguishing

4 Field situation during the 1986 field research at Schöneck-Kilianstädten

among them is the separation of fossil associations which may lead to better understanding of the composition and distribution of ancient plant communities in space and time.

The material from the surroundings of Winnweiler (Palatinate) was mainly collected on ploughed fields (Fig. 3) which have repeatedly been visited over the years by the first author. There are numerous small specimens in the respective collections, but some pieces of considerable size have also been recovered. The locality of Schöneck-Kilianstädten (Wetterau) was discovered by amateur archaeologists when a building site was being prepared for subsequent construction (Fig. 4). Systematic excavations by the Palaeobotanical Section of the Forschungsinstitut Senckenberg have yielded a large amount of material including a

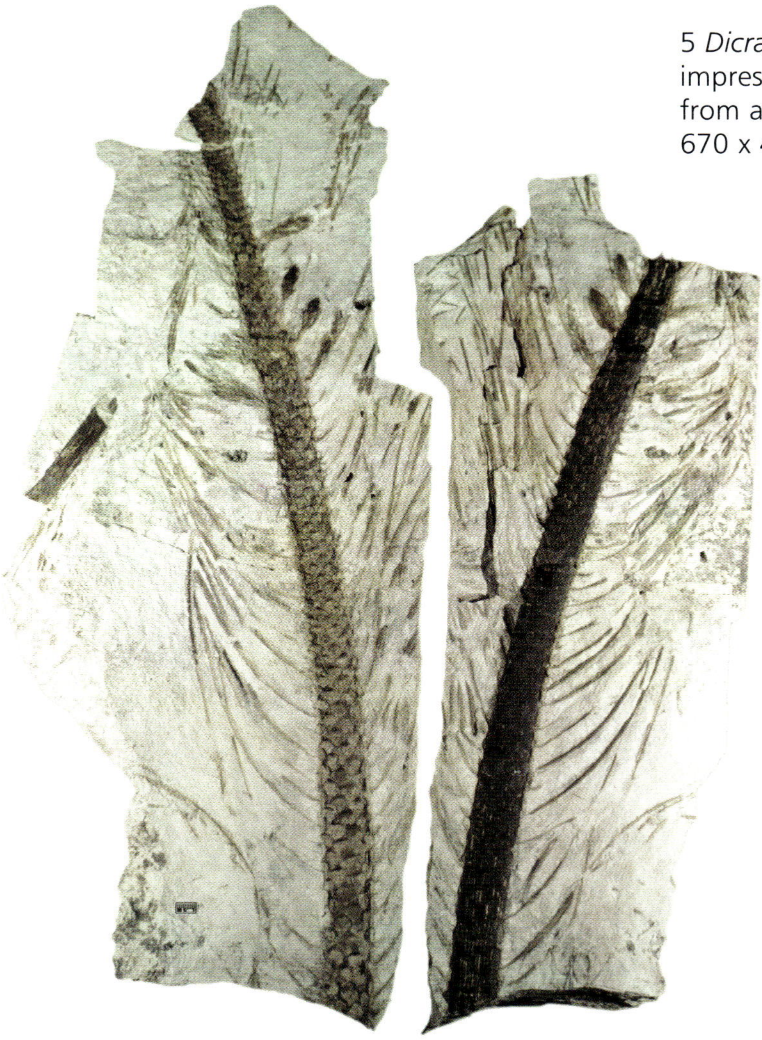

5 *Dicranophyllum hallei*
impression of a herbaceous conifer
from ash layer number 3 near Winnweiler
670 x 440 mm

6 Dead conifer trees partly entombed by pyroclastic deposits from the 1980 eruption of Mount St. Helens
situation in September 1999

number of individual stems. Studying these new and extremely well-preserved materials has led to results which have shed new light on the anatomical structure of early conifers.

Ash, Debris, and Mud

It has long been known that localities with petrified wood are regularly associated with deposits of volcanic origin. Burial by volcanic material is generally the best chance for preservation of the hinterland vegetation which would otherwise hardly be preserved. About 280 million years ago, in Rotliegend times, the area which today forms the Palatinate was characterized by active volcanism. The resulting landscape varied in elevation and was consequently covered by a mosaic of different plant associations. Volcanic ash was washed into depressions where it formed layers of considerable thickness. There is a twofold advantage of such volcaniclastics for the preservation of plants. The respective material is rich in siliceous components which are comparatively easily dissolved by percolating water. Precipitation of silica from these solutions initially takes place especially along the durable cell walls of the embedded plant fragments and stabilizes their framework. The remaining empty space within the individual cells is subsequently permineralized.
On the other hand, such fine-grained deposits form detailed impressions. The localities of the Wetterau were obviously more distant from the volcanism, but the respective strata still include enough volcanogenic material which was transported over some distance. Because both areas of deposition (Palatinate and Wetterau) were somehow connected in Lower Permian times, an almost contemporaneous formation of the petrifactions seems possible but cannot yet been proven unequivocally.
In both cases, flows of mud and debris swept trees down from higher elevations. A similar situation (Fig. 6)

7 Gymnospermous stem of the *Dadoxylon* type in transverse section
Schallodenbach, 480 x 310 mm

was observed during the 1980 eruption of Mount St. Helens (Washington, U.S.A.). It should also be mentioned that our silicified material is always accompanied by impressions of other plant remains.

Only *Dadoxylon* ?

Isn't it true that most collectors would like to imagine exotic trees from the distant past as the sources of their petrifactions?
These aficionados are likely to be disappointed by experts who, when asked for their opinions about the specimen's identity, all too frequently say, "It's just *Dadoxylon*".
But what is *Dadoxylon* really?
The word, in fact, is merely a scientific name for an artificial entity comprising wood remains of a common anatomical structure which is today generally restricted to araucarias, a group of ancient conifers currently limited to the Southern Hemisphere. This type of wood was not limited to the early conifers in Permian times, but was also developed by cordaites, an extinct group of plants with some systematic relationships to conifers. It has long been thought that the wood of cordaites and conifers could not be distinguished from one another without organic connection to pith and primary structures.
Dadoxylon-type wood is definitely the most frequent type of petrified wood throughout the entire Upper Palaeozoic although it may be difficult to decide whether the wood of a particular specimen is derived from a cordaite or a conifer. The density of its structure is another reason for the frequency of *Dadoxylon* specimens. This would have undoubtedly resulted in their being highly resistant to mechanical disruption and decay.
Many of the petrified logs were definitely not preserved *in situ*, but represent instead driftwood that was not permineralized until after it had finally settled.

8 Historic specimen of a coniferous stem with pith and remains of surrounding xylem
This specimen was found at Tholey and studied for the first time by Weiss and Potonié at the end of the 19th century.
Collection of the Museum für Naturkunde Berlin
103 x 69 mm

New Specimens – New Results

Tylodendron, Casts of Pith Cavities

Tylodendron is likewise a scientific name which was given to natural casts of the pith cavity of early conifers. Such specimens are characterized by a rhomboidal pattern of surface furrows which correspond to impressions of conducting strands on the internal side of the wood. Nodal branching of the plants is reflected by a regularly repeated increase in the diameter of the respective casts.

Internal Structures of the Pith

Regularly distributed, plate-like masses of sclerenchymatous cells, known as sclerenchyma or sclerotic nests or plates and occurring within the pith (Figs. 9-11), are characteristic of conifers but unknown among cordaites.

The pith of conifers remains closed for some time and an irregular transverse chambering develops late. Chambering is caused by longitudinal growth and may be similar to the conditions in cordaites, where it

10 Minor coniferous axis with transversely splotched pith Winnweiler, 40 x 30 mm

9 Longitudinal section of a coniferous stem with increasing diameter of the pith at a node, Kilianstädten 175 x 97 mm

11 Individual sclerotic nest with thick-walled cells, 2.4 x 1.6 mm

13 Thin section of a special type of wood from Kilianstädten showing secondary tissues within the pith, 50 x 47 mm

12 *Artisia*-like chambered pith of a coniferous stem from Kilianstädten 30 x 34 mm

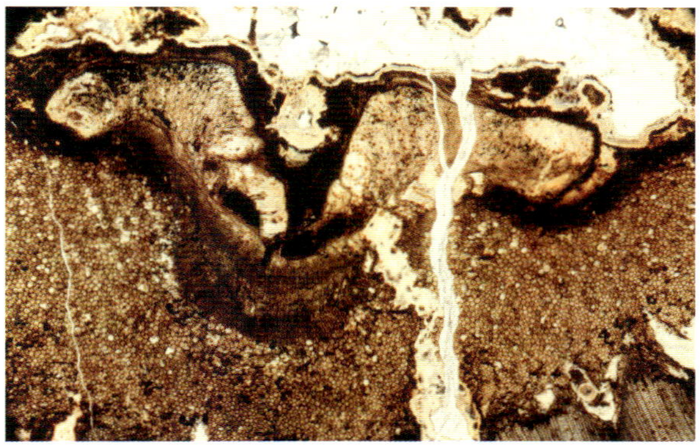

13a Detail with plate-like complex of secondary tissues within the pith, 6 x 4 mm

forms earlier. This has surely caused some past confusion between the respective fossils of cordaites and conifers.

In addition to sclerenchymatous or sclerotic plates, a number of coniferous axes from the Palatinate and the Wetterau show plate-like complexes of secondary tissues on transverse sections of their pith (Fig. 13). These obviously occur preferentially at the nodes, but their formation and possible function remain unknown. They provide evidence of the remaining potential of the tissue within the pith for subsequent cell division and differentiation.

Initial Growth

The conducting strands which are first formed during the initial growth of the young shoots are thus called "primary strands" or "primary xylem". In transverse sections of coniferous axes, they form wedge-shaped centripetal protrusions (Fig. 14). These represent the first-formed wood and the transition from the pith to the secondary xylem or wood. These protrusions are responsible for the characteristic pattern of furrows on the fossil casts of pith cavities which have been called *Tylodendron*. The distance between individual primary strands distinguishes two distinct types of *Tylodendron* in our material. A radially oriented, longitudinal section of the primary xylem (Fig. 15) shows ring-like or spiral thickenings of the cell walls of the individual conducting elements (tracheids). The thickenings prevented rupturing of the tracheids during the rapid longitudinal growth of the young shoots. The tracheids that formed first within the primary strands are called "protoxylem" and can be recognized by their having the smallest diameter. The second-formed part of the primary xylem is called "metaxylem". In cross-section, its tracheids are larger in diameter than those of the protoxylem and randomly distributed as compared to the radial alignment of the tracheids of the secondary xylem.

The Wood Itself

The longitudinally elongated conducting elements (tracheids) of the wood proper are formed during secondary growth of the stems in woody plants and thus form the so-called "secondary xylem" (Fig. 16).

14 Transition from wedge-shaped primary xylem to the pith in transverse section
coniferous stem from Winnweiler
3.6 x 2.4 mm

16 Polished transverse section of wood from a coniferous stem, Winnweiler, 1.5 x 1 mm

15 Radial section of the primary xylem of a conifer from Winnweiler; individual tracheids at the transition to the pith with spiral thickenings
0.75 x 0.5 mm

17 Polished radial longitudinal section of a specimen of wood from Winnweiler; individual tracheids show uniseriate to biseriate pitting
1.5 x 1 mm

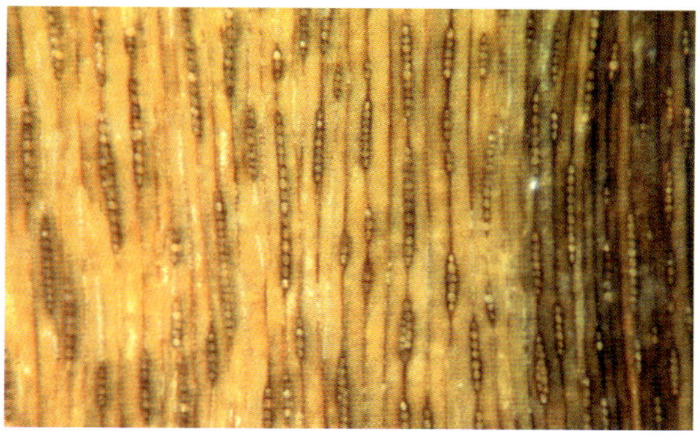

18 Polished tangential section of the secondary xylem of a coniferous stem from Winnweiler; rays are evident as vertical rows of individual cells
1.5 x 1 mm

20 Tangential section of a coniferous stem from Winnweiler with leaf traces marked
65 x 32 mm

19 Imprint of a coniferous axis in pyroclastic deposits from Chemnitz with scars of branches and leaves
Collection of the Museum für Naturkunde Chemnitz
diameter of the axis 12 mm

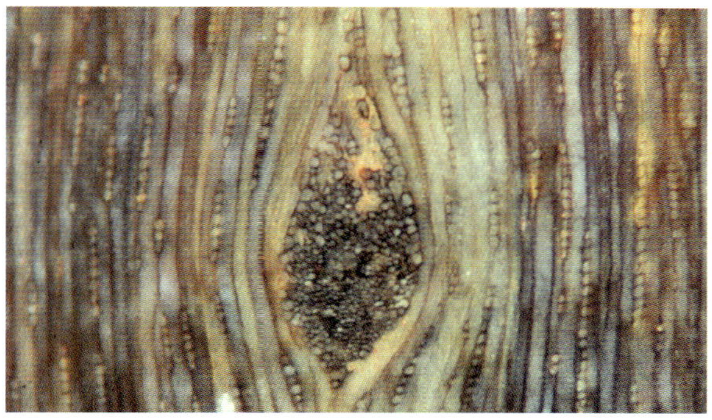

20a Individual leaf trace of a conifer, tangential section
Winnweiler, 1.5 x 1 mm

In transverse section, they are radially aligned between the rays. A radial longitudinal section of the wood (Fig. 17) shows pores or "pits" that connect laterally between individual tracheids. They are responsible for the continuous transport of water and nutrients through the tracheids of the stem. Pitting of conifers is arranged in one or two, in rare cases three, rows of pits per tracheid.
A tangential longitudinal section of the wood (Fig. 18) clearly displays the arrangement and character of the rays.

Leaf Traces and Branching

Branching of the early conifers from the Permian was arranged in whorls similar to that seen on extant araucarias. Accordingly, ultimate branching systems were functionally organized like fern fronds with a penultimate branch in a rachis-like position and ultimate branches equivalent to the pinnae. Similar to

21 Section of a wood from Neubamberg
showing false "annual rings"
300 x 230 mm

21a Such false "annual rings"
are caused by deformation
due to compaction
6 x 4 mm

22 Detail of a conifer twig
showing growth rings
6 x 4 mm

extant araucarias, the branches were covered by spirally arranged (Figs. 19-20) and more or less hooked, needle-like leaves with a single bundle.
The branching mode of the Permian conifers is well exposed in the imprint of an axis embedded in a volcanic mudflow and may be directly compared to the conditions in extant araucarias. Simple conducting strands (traces) which originally connected individual leaves to the primary strands can be observed in tangential sections of the fossil wood. According to the arrangement of the leaves, traces were spirally arranged in the early conifers.

Annual Rings or Not ?

Our material frequently shows widely spaced growth rings with a more or less irregular distribution on transverse sections (Fig. 22). Sometimes they are only weakly expressed, sometimes well developed. This has repeatedly led to discussions about seasonality and the character of the climate.
But caution is needed here because ring-like structures in fossil wood may sometimes be misinterpreted. They may even represent a zone of cells which collapsed due to pressure caused by compaction of the embedding sediment (Fig. 21).

Unequal distribution of internal and external stress during growth may have caused incomplete growth rings. In other cases, irregularly distributed but more or less well developed and circumferent zones of cells with thickened walls and a comparatively small diameter may reflect irregularly occurring periods of unfavorable growth conditions caused by drought, fire, or ash cover due to nearby volcanic eruptions. We cannot, however, fully exclude the possibility that true annual growth rings developed in localities with a more fully expressed seasonal microclimate.

23 Burnt stem from Winnweiler with a characteristic shrinkage pattern 115 x 85 mm

24 A burnt stem with preservation of the central part, Winnweiler, 280 x 260 mm

Traces of Fire

As is to be expected in areas with active volcanism, charred wood is regularly observed in our material from the Palatinate. The surfaces of the affected logs show a characteristic blocky pattern which was caused by shrinkage due to heat and fire (Fig. 23). Despite surface charring, the central part of larger logs is frequently well preserved. When cut, charred wood often shows an attractive pattern of black, angular pieces of charcoal separated by crevices filled with bright red agate (Fig. 24) which aesthetically enhance an impression of fire.

Rarely Preserved

Almost all of the large petrified logs show no traces of bark because they were abraded during transport. There are, however, a few rare fragments of twigs from the Palatinate with remains of bark. One of the specimens even shows the meristematic layer (cambium) which formed the secondary xylem (Fig. 25).

Subterranean Parts

Woody roots (Fig. 26) have been recognized among the petrifactions from both the Palatinate and the Wetterau areas. Known as *Amyelon*, they are characterized by the lack of a

25 Branch of a conifer in cross-section with cortical tissues preserved, Winnweiler 13 mm diameter

26 Cross section of a root from Winnweiler; with two primary strands separated by a cleft 20 x 14 mm

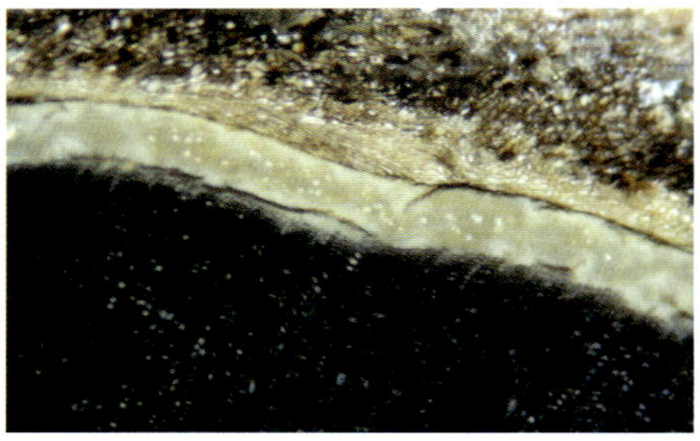

25a Transition from secondary xylem to cortical tissues with cambium between xylem and phloem, 2.4 x 1.6 mm

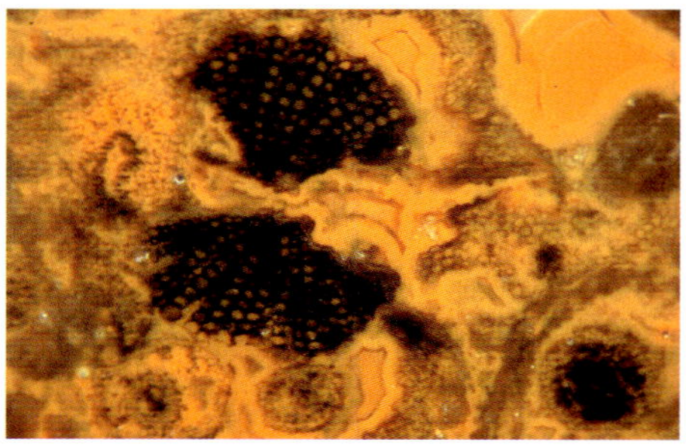

27a Detail with cross-section of a single root showing two primary strands, 3.6 x 2.4 mm

27 Permineralized soil penetrated by many roots Schallodenbach, 80 x 50 mm

28 Fragment of a root with imprint of a climbing fern and fern roots in the decayed interior, Winnweiler 50 mm diameter

29 Longitudinal section of petrified wood from Winnweiler with circular galleries, 54 x 36 mm

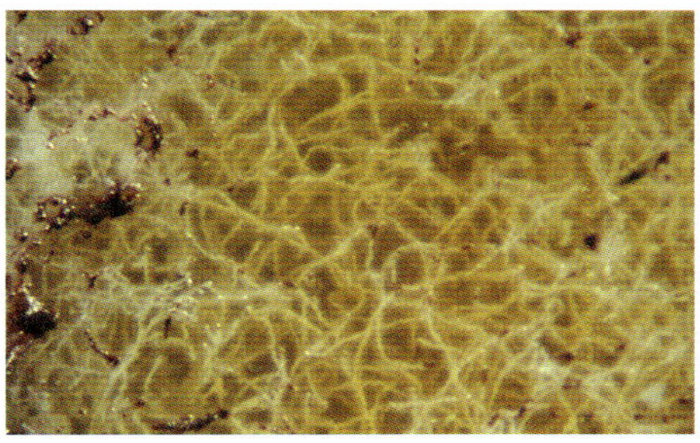

28a Fungal hyphae in the center of the root 1.5 x 1 mm

28b Fern root with primary and cortical tissues 2.4 x 1.6 mm

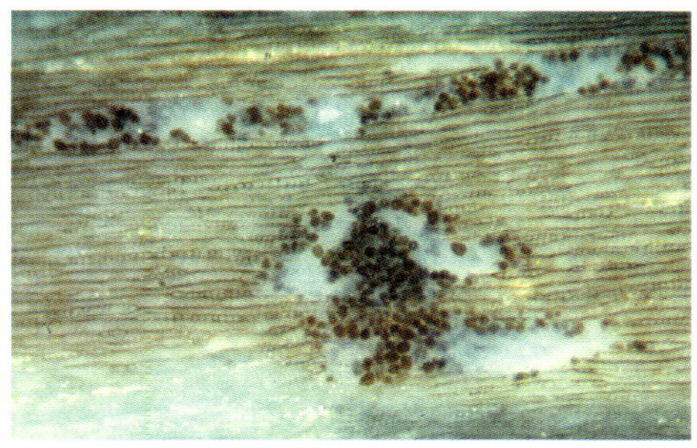

30 Detail from the margin of a decayed branch with galleries filled by coprolites which were probably produced by mites Schallodenbach, 2.4 x 1.6 mm

pith in combination with a central pair of primary tissues. However, it still remains impossible to distinguish between cordaites and conifers in isolated roots. In addition to isolated fragments of woody roots, permineralized soil horizons have been found in the Palatinate (Fig. 27). These are riddled with anatomically well-preserved fine roots which are obviously of gymnosperm origin, but their exact affinities remain unclear.

Wood Parasites

In a few specimens, the pith appears to have been weakened or even more or less completely destroyed. Fungal hyphae can regularly be observed in the pith cavity of these specimens. A few woody roots from the Palatinate have a central cavity with fungal hyphae and roots of ferns (Fig. 28). The outside of one of these specimens clearly shows invasion by the fern. Most probably, the respective axes had been decomposed from the center by fungi. Fungal decomposition liberated nutrients which could have been used by the fern.

Animals That Fed on Wood

Living and dead wood is frequently attacked by a number of animals, most of them insects and other arthropods. It is thus not surprising that some of our specimens of per-

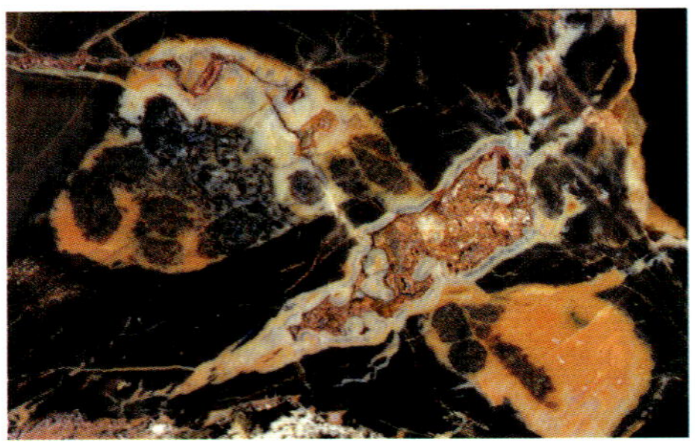

31 Wood with large galleries
containing individual coprolites
Winnweiler, 36 x 24 mm

32 Longitudinal section of a subordinate
cordaitean axis from Schallodenbach
showing collapsed internal diaphragms
43 x 42 mm

mineralized wood show various types of feeding traces (Figs. 29-31). Distinct galleries which are circular in cross-section may be distinguished from irregular cavities, both with feces of the feeding animals preserved almost in situ. The size of these coprolites may be characteristic of the organisms that produced them. Large amounts of tiny coprolites which are regularly associated with the irregular cavities can most probably be ascribed to wood-decomposing mites. The larger galleries and coprolites could have been caused by insects or larvae of insects. If this is true, then this would be one of the oldest evidence for the existence of wood-boring insects or larvae of insects.

Rare *Cordaites*

The natural casts of the pith cavity of the extinct cordaites, called *Artisia*, are characterized by a regular and distinct transverse striation which reflects the original chambering of the pith (Figs. 32-33). Growth of the pith did not keep pace with the elongation of the growing stems, which ruptured at intervals and left a diaphragm of tissue. Eventually these diaphragms disintegrated and left a more or less continuous hollow cylinder. When the plant died, the central cavity sometimes filled with sediment and a cast was formed with remnants of the diaphragms causing the *Artisia*-pattern on the surface. In contrast to *Tylodendron*, *Artisia* casts do not show a distinct longitudinal striation because the irregular primary xylem of cordaites scarcely formed wedges. Also, leaf traces of cordaites occur in pairs and are larger than those of conifers. Tracheids belonging to the wood of cordaites commonly show up to 5 rows of pits (Fig. 34) which form a quite distinctive honeycomb pattern by flattening the borders of each other due to close contact. Among wood of extant plants, this type of pitting is limited to the araucarias and thus commonly referred to as "araucarioid pitting".

33 *Artisia* (imprint and natural cast) from Winnweiler
120 x 70 mm

34 Radial longitudinal section of
corda tean wood from Winnweiler
showing multiseriate pitting
1.5 x 1 mm

Even Leaves are Petrified
Cross sections of permineralized pecopterid pinnules which were found in distinct layers in the Rotliegend of Saxony (Barthel, in this volume) have for some time been misinterpreted as representing the larvae of insects. The respective pieces of rock were therefore called "*Madensteine*" (maggot stones) in the 19th century.
Similarly preserved leaf remains have now also been discovered in the Palatinate, but they clearly belong to pteridosperm foliage of the *Callipteris naumannii* type (Fig. 35).

35 Chert with permineralized pteridospermalean leaf remains of the *Autunia naumannii* type
Schallodenbach locality

a - pinnules on the chert surface
width 36 x 24 mm
b - transverse section of a pinnule
width 2.4 x 1.6 mm

RONNY RÖßLER

Zwischen kostbarem Erbe und eigenem Erleben - Paläobotanische Forschungen im Perm von Chemnitz, Deutschland

Between Precious Inheritance and Immediate Experience - Palaobotanical Research From the Permian of Chemnitz, Germany

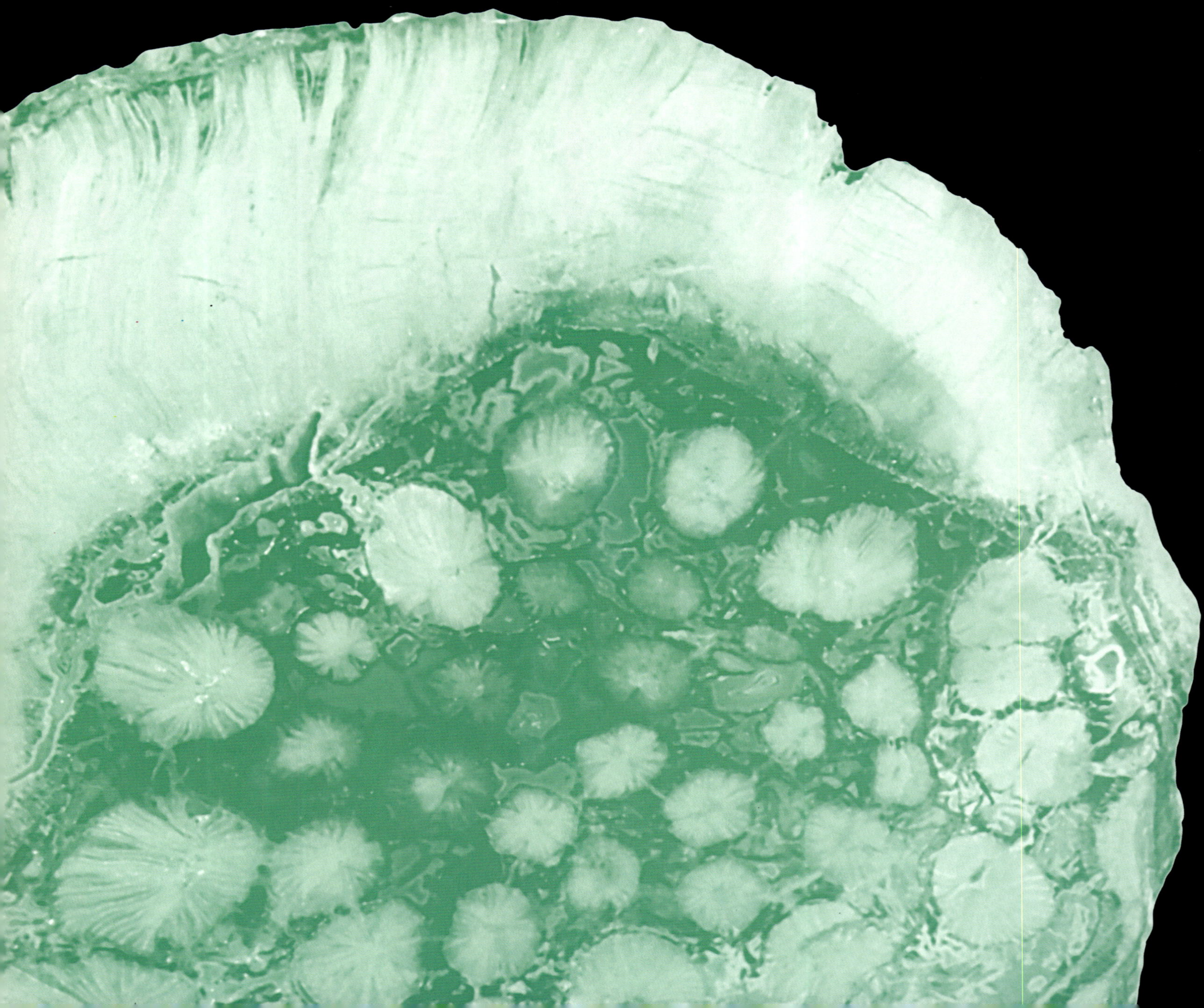

Not only has the so-called "petrified forest" in Chemnitz provided numerous historical original specimens, it has also made possible new discoveries in recent years. In addition to this, the comprehensive collection in the Chemnitz Museum of Natural History serves as a source of unique material with which new discoveries can be compared. Between these historical and contemporary contexts, several groups of plants deserve – and receive – somewhat closer examination here.

Ferns: Growth Form Diversity

Let's begin with the stately tree ferns, plants which formed part of the vegetation in widely diverse habitats ever since the Upper Carboniferous, and whose majestic appearance characterizes the moist cloud forests of the Southern Hemisphere. From the remote past and into the present day, tree ferns have preserved their principle of construction, which relies on a dense mantle of countless aerial roots as a supportive base around a stem containing numerous vascular bundles having branching fronds attached. The largest of these plants, the Permian Psaroniales, raised their finely divided fronds to heights of as much as 15 m without secondary reinforcement of their vascular tissue. Some of these early palaeophytic giants (especially the genus *Psaronius* Cotta 1832) had extensive palaeogeographic and stratigraphic distribution. Nearly contemporary with these are the *Psaronius* discoveries from the Euro-American region, China, and South America. From a stratigraphic point of view, documented occurrences range from the Carboniferous in the Variscan Foreland Basin up to the Triassic. Their various species grew in moist riverine lowlands, on peat-forming forested mires, and even at intermittently more drier sites. In many cases, however, only one or two species dominated a particular site. The most impressive fossil specimens of the psaronians are petrified trunks, but we also find examples of these plants complete with their petioles, pinnate fronds, and fructifications, preserved in "coal balls" (Rothwell, in

1 Terminal portion of a *Psaronius* trunk showing leaf scars preserved in tuff
photo: Gerhardt
height 158 cm

2/2a *Psaronius* sp.
transverse section of a beautifully colored mineralized trunk with numerous epiphytes of *Tubicaulis* cf. *berthieri*
photo: Gerhardt, 56 cm diameter

detail nearly 8 x

this volume) or as imprints in accompanying sediments. First identified as silicified segments of trunks by B. Cotta in 1832, the term *Psaronius* is currently used to describe the entire plant. Psaronias had erect, non-branching trunks and achieved a basal diameter of as much 1.5 m. A trunk segment with the largest diameter found thus far was described by Sterzel in 1887, who named it *P. weberi* (Fig. 3) in honor of its finder. The longest *Psaronius* trunk, which is unfortunately only a segment of a trunk with neither its base nor apex, likewise comes from Chemnitz and was reported to be 5.27 m long. Longer segments of trunks repeatedly show that the diameter remains nearly constant over many meters in length; only the base of the trunk displays a noticeable thickening in its mantle of roots. Inside the apparent trunk, the number of vascular bundles and the complexity of the entire construction both increase with the height of the trunk. The vascular bundles at the base of the trunk are surrounded by a thick armour of adventitious roots. An inner zone where the roots are embedded within a thin-walled, secondarily formed, cortical tissue can be distinguished from an exterior mantle. Towards the outside, the roots are freestanding and provide suitable substrate for a variety of other plants, for example, various climbing and epiphytic plants

(Fig. 2). Large specimens of *Psaronius* having rounded-elliptical petiole scars on the upper stem's surfaces are very rare (Fig. 1). One such specimen was found by the collector and independent scholar Otto Weber of Chemnitz, whose unique collection is now owned by the Chemnitz Museum of Natural History.

The history of Psaroniales research begins in the early 19th century, which means these ancient plants were avidly sought objects of research in the days when the science of palaeobotany was still in its infancy. The history of their collecting, however, is much older than the history of their research. As early as 1740, people had already begun to discuss and avidly collect these objects, which were often described simply as "precious stones". Having become filled during the course of its mineralization with colorful layers of chalcedony, each root appears as a particolored little agate and clearly shows why these pretty fossils were so attractive to collectors. This is also why these fossils became a coveted raw material for jewelry manufactured during the 17th and 18th centuries (Fig. 4). The first written mention of

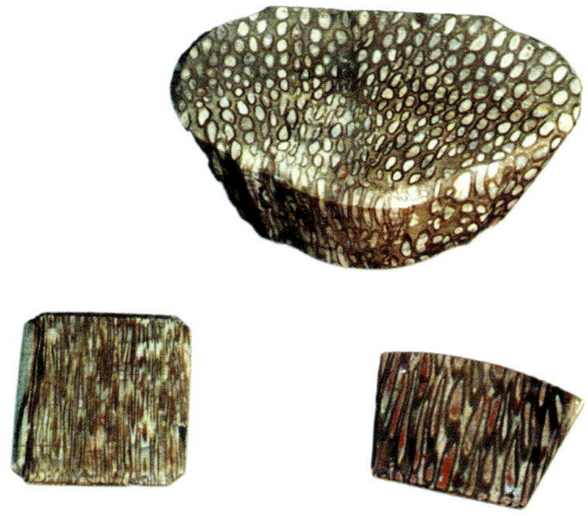

4 Pieces of jewelry made from *Psaronius* from the collection of Duke Anton Ulrich of Meiningen (1687-1763), Naturhistorisches Museum Schleusingen, photo: Noll

3 *Psaronius weberi* transverse section of a large trunk photo: Gerhardt, 62 cm diameter

the so-called "starry stones" was penned around 1760 by a physician and naturalist named C.F. Schulze. A sculptor named Vogel inserted an advertisement in a newspaper of 1804 in which he announced that he had "petrified wood from Chemnitz" for sale. Vogel mistakenly assumed that the starry stones among his merchandise were derived from the wood of petrified palm trees. A. Sprengel, who examined starry stones under a magnifying lens in the early 19th century, was the first to recognize the ferny nature of these specimens. During the mid-19th century, botanist A.J. Corda from Prague and palaeontologist C.G. Stenzel from Breslau devoted particular attention to the psaronias. The materials which they scrutinized were pebbles from classical Permian sites in Chemnitz and Nová Paka, as well as Upper Carboniferous tuffs from Bohemia. During the latter course of the 19th century, these finds were augmented by specimens found at Grand'Croix and Autun, France, and by specimens from South America. The second half of the 20th century brought a powerful boost in knowledge, especially of the psaronias, thanks to research conducted on specimens from the coal ball sites of North America's mid-continental coal basin. Nowadays, we include in the Psaroniales group not only *Psaronius*, but also the genera *Tietea* and *Tuvichapteris* from the Permian Southern Hemisphere.

Closely intergrown with the psaronias, we also find various epiphytic or climbing ferns. *Ankyropteris brongniartii*, a vine-like climbing plant which was widely distributed during the Permo-carboniferous, has been found in Autun, Nová Paka, and Chemnitz, as well as in the coal basins of England and North America. The slender climbing axes, which are spirally studded with small leaf scars, twined their way skywards primarily along the trunks of psaronias and calamites. Having had many branches, individual examples

5/6 Vine-like climbing fern *Ankyropteris brongniartii* eight axes on a *Psaronius*, outer surface and transverse section photo: Gerhardt, height 16 cm

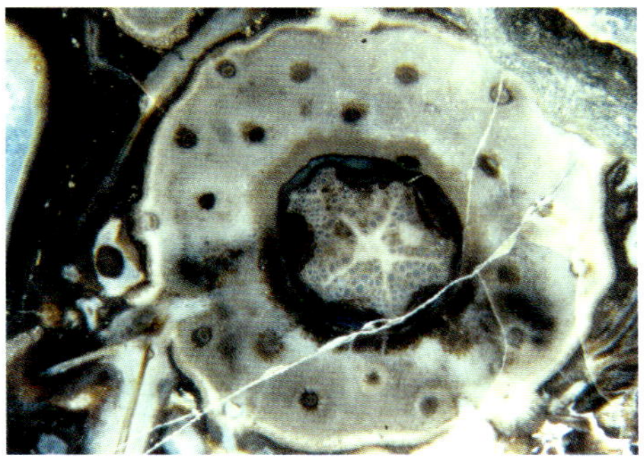

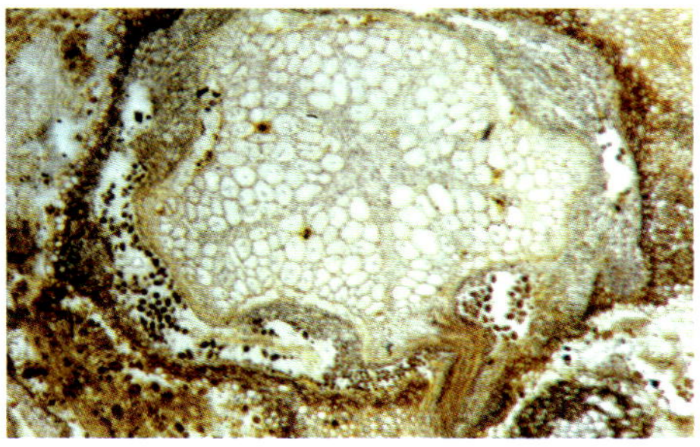

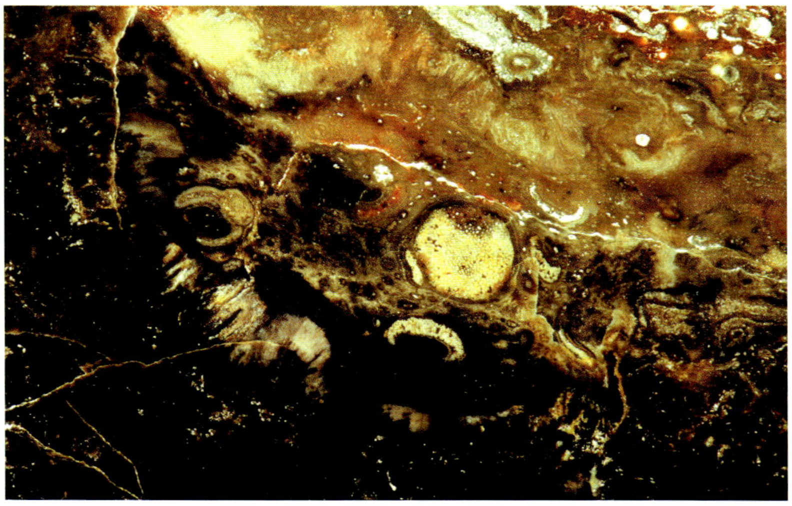

7 *Ankyropteris brongniartii* transverse section, Nestler collection, 5 x

8 Feeding tunnels filled with small coprolites in the phloem area of a climbing axis of *Ankyropteris brongniartii*, 13 x

9 *Tubicaulis* cf. *berthieri* epiphyte in outer root mantle of *Psaronius,* 4 x

10 *Tubicaulis* cf. *berthieri* epiphyte in outer root mantle of *Psaronius,* 3 x

show up to 8 climbing axes which surround the tree fern in a dense network (Figs. 5-6) or which have grown into the tree fern's basal root mantle. A dense mat of small aerial roots accompanies the climbing axes and anchors them to the trees. Attached petioles show the typical H-shaped vascular bundle in the form of a double anchor from which the plant derives its name *Ankyropteris*. Stenzel based his studies on some finds from Nová Paka, although the plant had already been named and described from the Permian of Autun in 1869 by Renault as *Zygopteris brongniartii*. Numerous highly impressive pieces exist from Chemnitz, and these specimens outstandingly reveal the typical growth form of this fern. Although initial hints exist to suggest that *A. brongniartii* axes belong to *Senftenbergia* fronds and sporangia, the final proof remains one of the secrets from which this book derives its name. We do know for certain, however, that this plant must have been a very flavorful one – at least as far as the taste buds of oribatid mites are concerned. In 1998 and thereafter, it was shown without a doubt that the phloem areas of this plant, as well as other basic tissues, exhibit evidence of insect feeding in the form of coprolite-filled tunnels (Fig. 8). The outer root mantle on psaronians were penetrated by another, epiphytically growing fern. According to a note in the literature by Indian palaeobotanist B. Sahni, numerous small epiphytes of *Tubicaulis* cf. *berthieri* (Figs. 9-10) had been identified in psaronias from the Chemnitz collection. The first mention of such epiphytes, however, came from France: C.E. and P. Bertrand published a report in 1911 about a small Perm an *Tubicaulis* from Autun which they found unlabeled in the Renault collection in Paris and which they then named *T. berthieri*. Later, around the middle of the 20th century, S.H. Mamay published a report

11 *Asterochlaena laxa*
Transverse section of a small actinostelic tree fern stem
original size

12 *Asterochlaena laxa*
longitudinal section of a small stem
with departing petioles
Güldner collection, original size

about a unique find of these plants in Illinois. Additional discoveries followed from American coal balls and the pebbles of Grand'Croix, France. The crowning moment in the history of these discoveries occurred with the unearthing of more than 40 of these little epiphytes, on and in psaronias from Chemnitz during the last years. The fact that these epiphytic and climbing plants existed in much greater numbers than previously known provides an important detail that will support the reconstruction of the Permian vegetation.

Numerous smaller tree ferns, e.g. the extremely rare or unique *Asterochlaena* or *Grammatopteris*, have always stood more or less in the shadow of the more numerous and more commonly found psaronias. These smaller plants, however, serve as impressive examples of mosaic evolution. They show manifold combinations of diverse types of steles and petioles found. To get an idea of their

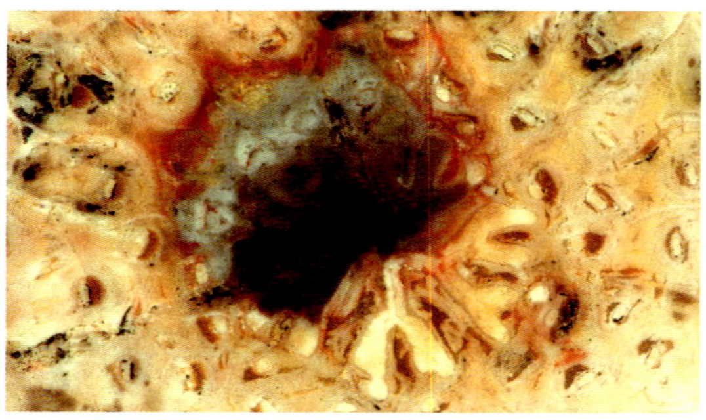

13 *Asterochlaena ramosa*
Actinostele and leaf traces, 2 x

14 *Blechnum brasiliense*, a small tree fern from the South American tropics
8-10 cm stem diameter

growth form and to more vividly imagine what they must have looked like based on the countless slices from individual specimens scattered throughout the world's museums and collections, we can compare them to the recent *Blechnum brasiliense* (Fig. 14). Only a few specimens exist of *Asterochlaena laxa*. According to Stenzel, who first named the genus in 1889, all but one (a lone specimen from Nová Paka) were found in Chemnitz. *A. laxa* has an ideally formed actinostele surrounded by loosely attached, spirally arranged petioles (Figs. 11-12). New discoveries of small tree ferns having actinosteles from Permian deposits in Brazil (Dernbach et al., in this volume) make the *Asterochlaena* discoveries unique comparative material.

The apothecary F. Nindel, a great aficionado and avid collector of petrified woods from Chemnitz, wrote a brief report in which he announced his discovery of the remains of "a silicified, herb-like fern stalk". This trunk, which measured 26 cm in length and nearly 7 cm in diameter, was unearthed during construction excavations in Chemnitz in 1915. This sensational specimen (Fig. 15) was immediately recognized as a valuable and correspondingly costly item. Financial support from patron of the arts and sciences, R. Baldauf, helped to provide the funds needed to purchase two segments from this specimen for the Saxon Mining Academy in Freiberg. Baldauf's generosity is duly praised in the first scientific report about this specimen, in which R. Beck names it *Protothamnopteris baldaufii*. The name of the genus alludes to the basal position which occupies it in the phylogeny of the royal ferns (Osmundaceae). *P. baldaufii* (now *Grammatopteris*) was a small tree fern with a basal root mantle and a stem similar to the oldest members of the Osmundaceae from the Upper Permian of Russia. The most comprehensive study on *G. baldaufii* was published by Sahni. The task of classifying this plant was facilitated by a find from Autun. Renault named this specimen *Grammatopteris rigollotii* in honor its collector in 1893. Valuable new discoveries of closely related forms from the Permian in China (*Rastropteris pinquanensis*), as well as from Tocantins, Brazil (*Grammatopteris freitasii*) confirm that, despite the rarity of its fossils, this genus must have enjoyed extensive palaeogeographic distribution.

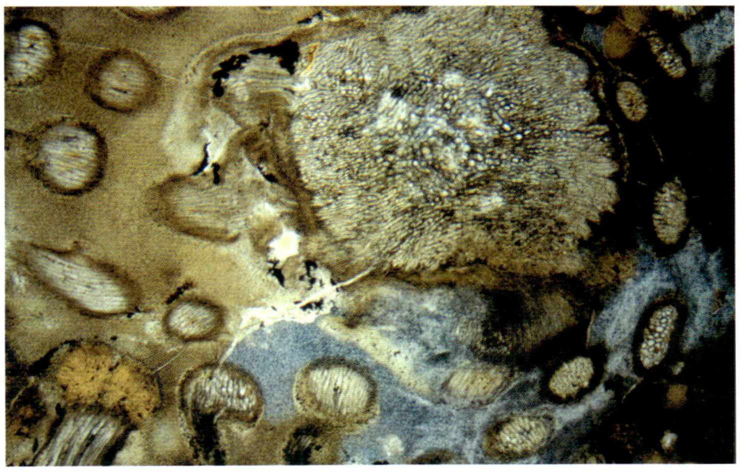

15a *Grammatopteris baldaufii*, detail of the protostele and leaf traces in transverse section, 8 x

15 *Grammatopteris baldaufii*, transverse section through the small tree fern, original size

A Few Pieces From the Puzzling World of the Seed Ferns

Whereas, especially in recent years, significant progress has been achieved in our understanding of a large number of climbing and herbaceous seed ferns, one of the most important members of this fossil group, the medullosas, still rank among the most enigmatic plant fossils. They were seed plants with fern-like leaves and ranked among the largest and most stately of the pteridosperms (Figs. 16-18). Their existence is the result of one of nature's most impressive experiments: the combination, optimi-

17 Star-shaped vascular bundle in the pith of *Medullosa stellata*, 5 x

16 Trunk segment of *Medullosa stellata* f. *typica* in longitudinal section
photo: Gerhardt, height 108 cm

zation, and adaptation of types of vascular bundles, growth forms, and reproductive strategies that occurred in response to the gradually changing environmental conditions that ensued towards the close of the Palaeozoic era. The cradle of their research was in Saxony, where a young scientist named B. Cotta was the first to describe *Medullosa*, basing this genus on material which had been found in Chemnitz and which belonged in his father's collection. Specimens from the Permian are also the last to be fossilized from this plant group before it became extinct. This extinction may have been due to the dwindling of the swamp habitats which had been the preferred dwelling places of the medullosas for more than 50 million years. Despite the effort of several generations of researchers, scientists are still wrestling with further elucidation and understanding of many aspects of these plants. No palaeobotanical textbook fails to mention the medullosas, although such books frequently include illustrations of allegedly "typical" plants which actually have little or nothing in common with Cotta's woody trunks that served as the generitype for this group of fossil plants. Impressive examples of medullosas from Chemnitz were collected by O. Weber, who made the thus-far most spectacular discovery in 1900. We can thank Weber's meticulousness as a collector for the discovery of nearly eight meter long trunks of *M. stellata*. He was very probably the most precise observer ever to have visited this rich site and locus typicus of the medullosas. Weber also discovered the attachment of the sometimes forked petioles of *Myeloxylon elegans* (Fig. 19) and *Medullosa leuckartii* trunks. Because he didn't know about the pteridosperms (which were not discovered until years later), Sterzel doubted the

18 Transverse section of *Medullosa stellata* showing more than 40 star-shaped vascular bundles in the central pith
33 cm diameter

19 *Myeloxylon elegans*
the petiole of *Medullosa leuckartii*
18 cm diameter

20 *Taeniopteris abnormis*
Leaf belongs to one kind of *Medullosa*
original size

21 *Medullosa stellata*
Tracheid rows from a so-called "plate-ring" at the periphery of the stem, height 45 mm

22 *Medullosa stellata* f. *lignosa*
Transverse section of a trunk with thick woody mantle and small pith
34 cm diameter

ferny nature of the leaf compressions, *Alethopteris* and *Neuropteris*, which had been associated with *Myeloxylon* in France, but had never exhibited any recognizable fructifications. Sterzel thought that the medullosas were an "intermediary group of plants with characteristics of both the cycads and ferns". In Chemnitz, pinnate fronds of *Alethopteris schneideri*, some of them with *Cyclopteris*, and *Taeniopteris abnormis* (Fig. 20) were found at the same place as *Medullosa* specimens and in adjacent tuff. The stems were usually slender, with periodical swellings, and they seldom attained diameters in excess of 10 cm. Only a few measured between 15 and 20 cm in diameter; and only a single one measured 48 cm across. The core of the stem is a broad, central pith containing one to as many as 40 lengthwise braided vascular bundles. Depending on the size and formation of these vascular segments, scientists describe them either as star or plate rings. The nearly total fusion of the peripheral plate rings creates a bulky wooden body. The taxonomy of the medullosas is still inadequately explored. In their day, Weber & Sterzel sought to understand the then-known species as a group of related forms. These two researchers emphasized the possibility that some of their proposed forms could perhaps be merely "... age, growth, or preservation variants of one and the same plant". Only the future discovery of longer sections of stems will provide us with the opportunity to identify various ontogenetic states of a single plant and to better understand the natural diversity of this genus. The group of forms according to Weber & Sterzel includes *M. stellata* Cotta 1832, to which the majority of Chemnitz medullosas belong, as well as *M. porosa* Cotta 1832, *M. solmsii* Schenk 1889, and *M. leuckartii* Göppert & Stenzel 1881. The most complete and most impressive specimen of *M.*

leuckartii was more than 1.5 m tall and had a diameter of 26 cm. This splendid specimen was found in 1910 by the Chemnitz resident, M. Güldner, a building contractor and collector of silicified woods. Because he carefully unearthed all of the fragments, including numerous petioles (some of which measured as much as 60 cm in length), it was possible to reconstruct the complete stem. Some of the later were found attached or close to the stem, thus proving that the petioles of *M. leuckartii* were spirally arranged at intervals of nearly 30 cm from one another and were not deciduous. The cross-section shows them to be obtusely triangular and very bulky. For the first time in 50 years, this specimen has only recently been displayed again at the museum in Chemnitz. Besides the impressive figure of the medullosa trees, many other herbaceous or climbing seed ferns are present. When one considers the speedy progress achieved in knowledge about these fossils during the past few years, and especially with regard to the exciting publications by M. Krings and H. Kerp from Münster, whose texts have called attention to adhesive discs, climber hooks, and other peculiar features of plant evolution during the Late Palaeophytic, one is tempted to view the tree-like medullosas not as the rule, but more nearly as the exception among the seed ferns.

The bushy to climbing *Callistophyton* plant still remains probably the best-known of the pteridosperms. Not only its growth form, its anatomy, and the evolution of its vegetative organs, but also the individual development of its seeds and microgametophytes have all been comprehensively studied (Rothwell, in this volume). Renault, who used the name *Poroxylon* to describe the first discoveries from Permian strata in Autun in 1879, identified them as remains of *Cordaites*. In 1999, we were able to prove the existence of *Callistophyton* roots in a dense plexus of basal *Psaronius* root mantles from Chemnitz. This explained R. Zeiller's 1888 illustration, which depicted a community (which had sometimes been described as rather doubtful) of *Dicksonites* fronds and *Psaronius* trunks in Upper Carboniferous material from Commentry, France and conclusively proved that *Callistophyton* indeed climbed on *Psaronius*.

The Woody Giant *Calamites* Trees

We're familiar with contemporary horsetails as herbaceous plants. Their erect aerial stems wither quickly, but their subterranean and extensively creeping rhizomes are exceedingly hardy – much to the chagrin of a great many contemporary gardeners. The principle of construction is very old, but during the Palaeozoic the aerial stems formed articulate, up to 10-m-tall, and often multiply branching axes around the woody mantle of a large pith. The calamites were undoubtedly perennial plants. Some of them contributed to coal formation, others formed groves along the

23 *Arthropitys bistriata*
median longitudinal section of a small main axis with diaphragms in the nodal area, original size

25 *Calamodendron striatum*
Detail of the woody stem
original size

24 *Calamodendron striatum,* transverse section of calamite with thick woody mantle around the pith original of Cotta (1832), British Museum of Natural History London, nearly original size

shores of rivers and lakes. Although they were able to spread by vegetative reproduction through their horizontal rhizomes, they also disseminated enormous quantities of spores. Fossil remains of the calamites number among the most common plant fossils from the Permocarboniferous. We usually find sedimentary fillings of the hollow piths, known as pith casts, which gave rise to the name "*Calamites*", or else impressions of the typical leaf whorls known as *Annularia* or *Asterophyllites*. Not only the outer form of the stems, but also their cellular tissue, have been preserved in the Permian of Chemnitz, as well as in Nová Paka and Autun. Although numerous specimens have been discovered, attached organs are very rare. Each form of preservation, the appearance of the surface of the fossil

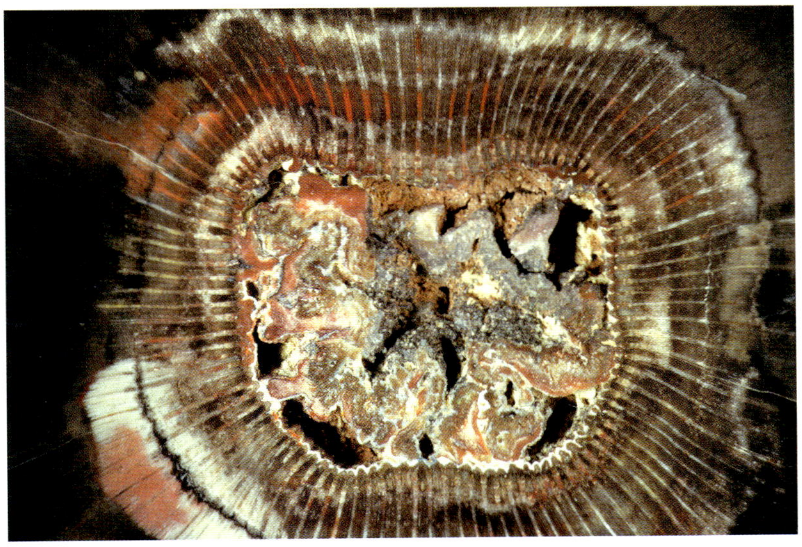

26 *Arthropitys bistriata*
Stem in transverse section with pith
width 45 mm

imprint, the internal anatomy of permineralization (Figs. 23-26), or the combined possibilities of fossil preservation in finely grained tuffs yield their own characteristic features, and the palaeobotanist's task is to bring these diverse aspects together in an effort to reconstruct these ancient plants. That this task is generally easier said than done was a fact well known to the young B. Cotta, when he confidently remarked in 1832 that, "perhaps a more practiced botanist will someday succeed in ornamenting these trunks with the leaves, blossoms, and fruits that originally belonged to them". We still know relatively little about the surface structures of the calamites. Some early collectors even removed the remains of vascular and cortical tissue from specimens that had been preserved as compressions. For many years, it only seemed worthwhile to keep and collect the pith casts, thousands of which were already filling collections. The study of calamite anatomy began in the early 19th century. B. Cotta explained the "Calamiteen" of Chemnitz as the, "... well-preserved structure of calamites", and thus distinguished himself as the first researcher to correctly recognize their place in plant classification. But because of its woody body, for many years Cotta's *Calamitea* genus was placed among the gymnosperms or even assigned to the *Sigillaria*. These authors interpreted *Calamodendron* Brongniart and *Arthropitys* Göppert as transitions between horsetails and conifers. These two important form genera for mineralized calamite axes are still commonly used today, and they are frequently encountered at Chemnitz, Nová Paka, and Autun. The principal difference between them lies in their anatomical features. In *Arthropitys*, the sectors of secondary wood alternates with large interfascicular rays. In *Calamodendron*, the rays are flanked by fibers which separate them from zones of secondary xylem (Fig. 25). In *Arthropitys*, a significantly variable amount of parenchyma occurs in the rays. The rays are variable in size and shape with each having one to four rows of parenchyma cells. Today we're certain that calamitean axes from the Permian suggest significantly more natural species than is currently included in their classification and that there were more growth forms than previously known. A truly comprehensive scientific treatment of calamite axes needs to be undertaken. Sensational new discoveries of Permian calamites from Tocantins, Brazil (Dernbach et al., in this volume) confirm that calamites had occurred successfully in moist habitats of the Upper Palaeozoic of the Southern Hemisphere. Perhaps the *Arthropitys* calamites will also prove to include forms with free-standing (without rhizomes) trunks such as had already been suggested in the "forest of Treuil" – a drawing, contained in an 1821 paper by Alexandre Brongniart, which depicts a quarry wall of Upper Carboniferous rock in the Loire Basin As far as the calamites of Chemnitz are concerned, we know that *Calamodendron striatum* is identical to *Calamites multiramis*, a plant that had a large, woody stem with closely spaced internodes and a crown of fronds similar to those found on tree ferns. Their connection to the leaves of *Annularia stellata* and strobili of the *Calamostachys tuberculata*-type had already been demonstrated from the Upper Carboniferous of Commentry by Renault in 1890. Numerous other discoveries in Lower Permian basins confirm that this plant was an important element of the forested mires. *Arthropitys bistriata*, with axes up to 20 cm in diameter and more than two meters in height, can be readily identified from the fossilized impressions in several specimens, including, but not limited to, the *Calamites undulatus* group with *C. alternans* Germar (preserved as a pith cast) and *C. varians* Sternberg var. *insignis* Weiss (the surface of which is preserved). Their branching indicates that they belong to *Calamitina*.
A whorl with 6 to 11 branches is

found in *Calamitina* at intervals of every 5 or 6 internodes. Purely morphological criteria, which were often used in the past as a means of distinguishing between different species, have proven to be highly variable and thus inadequate as criteria upon which to base taxonomic decisions. Among the calamites, the formation of secondary xylem might have led to a drastically skewed ratio between the plant's own mass and the capabilities of its vegetative structure. The typical calamites could only have flourished in environments characterized either by the lack of competition or by weak competition. Their Permian demise loomed ever closer, catalyzed by geotectonic changes which, in turn, influenced the global climate and created conditions in which new competitors could arise.

What Else Lived There ?

Our picture of the ancient habitat at Chemnitz is rounded out by a few faunal discoveries, for example, the remains of the skeleton of an *Eryops* amphibian. Without a doubt, we have every reason to expect that future finds will yield surprises because a rich arthropod fauna surely thrived there, as confirmed by the presence of coprolites and various types of feeding tunnels (Fig. 27) in the silicified woods. The reader can learn more about the fascination and the secrets of gymnosperm trunks of the *Dadoxylon* type elsewhere (Noll & Wilde, in this volume). These trunks are not only the most common, but also the largest in the "petrified forests" of the Permian era.

28 Jellyfish compressions in tuffitic sediments collection of the Freiberg University of Mining and Technology, original size

27 Feeding pattern in a piece of gymnosperm wood original size

W. WALTON WRIGHT

Die Triassische Chinle-Formation, U.S.A., und ihre fossilen Wälder

The Triassic Chinle Formation, U.S.A., and its Fossil Woods

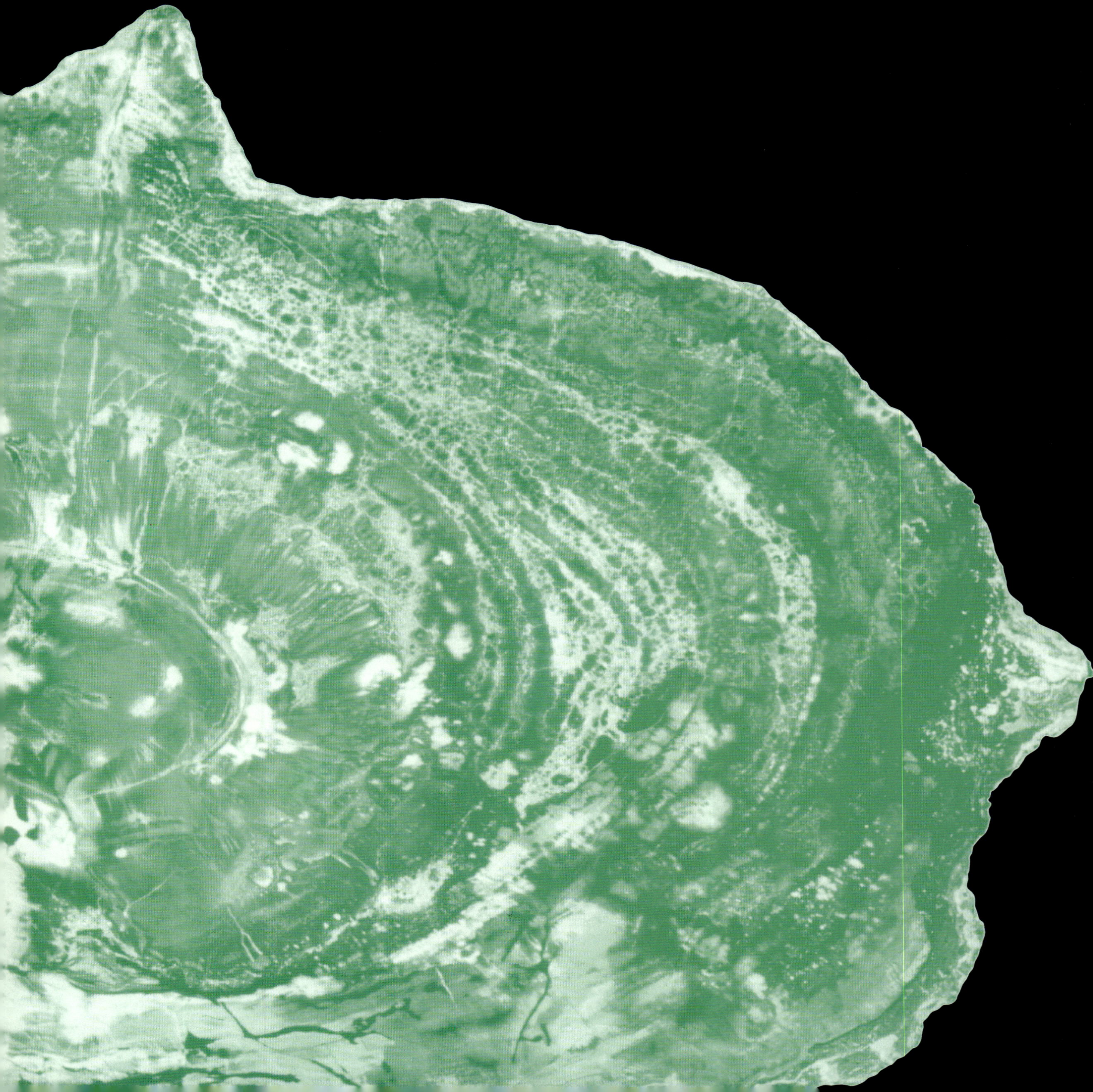

When people around the world think of petrified wood or petrified forests, the vision of logs in the Petrified Forest National Park of Arizona comes to mind. This is a spectacular national park, from a palaeobotanical point of view, but it is not the only locality where this scene of fossil wood and badlands of many colors can be experienced. As you look at the sections of the large logs turned to stone, don't simply think of them as interesting oddities of nature scattered like jacks. Try to visualize the petrified logs as a sampling of the trees standing in a tropical forest. In your imagination, look at the trees and see the forest, the forest ecosystem in an ancient setting near the equator. This fossil forest exhibits only a segment of an extensive wetland habitat that once extended over a swathe of land from Mexico, up through Texas, New Mexico, Arizona, Nevada, Utah, Colorado, Wyoming, and Idaho. This ancient landscape was an area of swamps, meandering rivers, lakes, flood plains, and deltas. It was a lowland with little topographic relief. It went from sea level in the west to a relatively low elevation highland in the east. The humid\wet tropical climate of the time provided ample year-round rainfall to sustain the large trees of an old growth forest. This long-past community of soils, microorganisms, invertebrate and vertebrate animals, and plants has been preserved in formations that were laid down in the late Triassic some 230 to about 205 million years before the present. This is the Chinle Formation, in its broad sense. It is the most extensive and complete section of continental sediments of late Triassic age anywhere around the world.

Time, one might think, is a straight forward/cut-and-dried fact. In a geologic sense, this is not the case. Geologic time is measured in million year increments, as years are measured in historic time. Each geologic time line and/or chart has its own version of where the million years before present break is between periods, epochs, ages, and stages. The names of the intervals of time are fairly consistent, but the absolute age is not. There are never radiometric dates for the out-

1-3 Logs from the Petrified Forest National Park

crop, bed, member, formation, or the lava flow, ash fall, you are interested in. The top of the Chinle is within the upper Triassic using one set of numbers, it extends into the lower Jurassic with another set.

But when did this story start? The beginning of the long version has its setting in the Pennsylvanian period of geologic time, some 320 million years ago in east central Utah, north to south across central Colorado and New Mexico, and west Texas into Mexico. At that time and place a mountain system was being pushed up. This ancient highland has been referred to as the Ancestral Rockies. These mountains were the drainage divide of southwestern Laurasia.

As these mountains were rising they were being weathered and eroded away, a process that was to continue for some 120 million years. We will pass over millions of years coming up to the early Triassic Moenkopi time, 247-240 million years ago. In this time, outward flowing streams off the uplift deposited clays, silts, and sands in tidal mud flats, alkaline flats, deltas and flood plains. These extensive flats blanketed the entire basin from the highland to the western coastline. This basin was flat with only a slight slope/gradient from the mountains to the sea. Thus, with only a slight change in sea level or land elevation, the shoreline fluctuated. Within the sediments were iron rich minerals that with subsequent weathering and oxidation would produce the deep red color staining the shales, mudstones, and sandstones of today's Moenkopi cliffs (Fig. 4).

Coordinated with the sea's retreat was the advance from the east of erosive forces. For the next 10 million years of the mid Triassic, the streams from the highlands dissected the marine flats, carrying away the fine muds, creating wide valleys and deep draws. This mid Triassic interval in time produced the unconformity between the early Triassic Moenkopi and the late Triassic Chinle Formations.

The initiation of rifting in mid Pangea (the beginning of the end of this supercontinent), transformed western Laurasia from a passive continental margin to an active margin. With the start of the westward continental plate movement, subduction of oceanic crust began. As the oceanic plate descended it was slowly heated, converted to a magma, which then rose melting its way upward through and incorporating material from the overlying margin of the continental plate. The magma plume upon erupting at the surface, produced rhyolitic volcanos. Thus on the southwestern edge of the continental borderland, a continental volcanic arc started to form. This would become an extensive volcanic highland which would contribute sediments and ash to the Chinle Basin for the next 20-25 million years. The volcanic arc, across the present Mojave Desert of southern California, continued through southern Arizona and into Sonora, Mexico. The Arizona/Sonora section of the volcanic arc constituted the Mogollon Highland (not to be confused with the present Mogollon Rim).

Vast and thick deposits of volcanic debris, ash to cobbles, were deposited in the upper Triassic Chinle. About 3/4 of the lower part of the Chinle is of this material, a volume of about 70,000 km^3. It is inferred, by this amount of volcanic debris, that volcanism was widespread in southern Arizona, but the provenance of this material, the volcanic centers from which it came, is unknown. The volcanism is thought to have been intermittent and occurred at widely separated localities. The southern Mogollon Highland has long been inferred to be the source of the volcanic debris. This inference has been derived from the structure of the Chinle deposits:

- sediments are coarser to the south and finer to the north
- clasts, cobbles to pebbles, within the sediments are more frequent in the south
- deposits are thicker to the south and thinner to the north
- stream patterns indicate a northerly flowing drainage system.

In the mountains of SW Arizona, there are volcanic flows and great thicknesses of clastic volcanic and sedimentary rocks of inferred early Mesozoic age. The Sil Nakya Hills west of Tucson are included in these volcanics. Rhyolitic volcanic material and activity is the critical ingredient and factor in the production of the Chinle deposits and the formation of the petrified wood.

As the rhyolitic ash buried forest remains, it was the water that aided in cutting off the oxygen, slowing decay. The fine texture and the frothy porous structure of the ash presented a tremendous surface area to the agents of weathering. The constituent minerals, upon weathering furnished the silica to the ground water system. Water moving in and around the tiny ash particles picked up, and carried dissolved silica into all of the cellular and intercellular spaces within the ancient coniferous trees, cycads, and ferns, encapsulating the organic matter, facilitating the preservation of these ancient plants for posterity. The ashes have weathered to clays, and water is now in the process of exhuming the long buried petrified forest for observation and examination.

The following is based upon my wan-

4 Red House Mesa, west of Blanding, Utah
In the foreground, the beveled upper slope of the Permian White Rim Member of the Cutler Formation stretches to the red cliffs of the lower Triassic Moenkopi Formation; sandwiched between the underlying Moenkopi and the overlying Wingate Formation cliffs and slopes of the upper Triassic Chinle Formation

5/6 Colorful badlands of bentonitic clay

7 Log with knots

dering along trails through stands of tall trees, and excursions by canoe down rivers bordered by tall trees in lowland tropical wet forests of Guatemala, Costa Rica, Panama, Columbia, and Ecuador; and by "armchair" exploring and climatic comparison in the Congo River drainage. During Chinle time, this region of the present western United States lay in tropical latitudes, within 15 degrees north of the equator.

As the present is the key to the past, the Chinle climate and habitat may be characterized by (1) year round temperature being fairly constant; (2) sun intensity being high at times during the day with little or no cloud cover; (3) temperatures being warm, only feeling hot for short times with little or no rain; (4) never having a real dry season, only a less wet season (particularly during Blue Mesa time as indicated by the fossil wood not showing growth rings); (5) rain possibly being seasonally heavy, maybe monsoonal, i.e. having several; (6) large storms during the summers, in the more northerly latitudinal areas; (7) surface stream flows being seasonally mediated by the moderating influence of the vegetation and the water stored in the soil of the drainage basin; (8) water level being relatively stable in the wetlands associated with but not connected to the streams and rivers; (9) water level in the rivers dropping, exposing sand and mud bars during short intervals without rain; (10) "wilting" of some plant species occurring during short rainless periods, "growth rings"/drought rings can be observed in some specimens of some species from Black Forest time; (11) rainless periods (less wet periods) being dryer, but the soils never dry; dryer does not mean arid.

Today, the Congo River basin, situated near the equator on the western side of the African continent, may represent the Chinle world. Though this river system, draining from the eastern highlands to the west coast is much more extensive, the climatic regime is analogous. In this setting there was and still is a tropical wet climate with a rain forest type vegetation, characteristic of the latitude. Further north from the equator a tropical monsoonal climate, in spite of the drier season, still supports the rain forest. The winds sweeping from the southwest off the equatorial ocean are the winds that, in Chinle time, distributed the volcanic ash.

The tropical rains bring the moisture and humidity, saturate the soil and feed the streams, and support the vegetation. The forest is dominated by trees with large diameters and great heights. This is a highly diverse community of species. The physical and morphological aspects of today's rain forest may replicate those of Chinle time. But, the present components of the vegetation did not exist in the upper Triassic - it would be 125 million years before monocotyledonous and dicotyledonous trees would come on the scene. Large gymnospermous trees with understory cycads and tree ferns, plus ferny herbaceous ground cover species, were sufficient to produce a Chinle version of a tropical wet to wet-drier rain forest.

Over the 25 years or so I have given talks, seminars, and led field trips, I have discussed the Chinle Formation in a broad sense, in which I include the Popo Agie Formation of Wyoming, the Dockum Formation of Texas, and the Dolores Formation of Colorado.

The Chinle Formation is named after Chinle, in the Chinle Valley, Apache Co., northeastern Arizona. Petrified wood occurs throughout the range of the Chinle, although it is not found uniformly through the formation. Some layers have large, long logs, some have small diameter logs, whereas others exhibit little or no petrified wood. As one looks at the long logs throughout the petrified forests of the Chinle, no limbs are seen along the trunks and few knots are found. Upturned, spreading root systems are not encountered, although some trunks with root bases can be found. The wood is preserved, but there is no bark. In many references it is stated that the logs of the forest came to their resting place as driftwood, carried down a river for perhaps 150 km or more. But, is it really reasonable to have all these logs as driftwood, and none in or near their place of growth? Probably not. Stumps were reported in growth position, but this was later discounted, rightly or wrongly. I collected a small stump, about 7 cm diameter and 17 cm tall, with 2-3 cm diameter roots extending radially to about 60 cm. The specimen appeared to be in growth position.

Where have all the limbs gone? In living araucarias and other trees of the rain forest, the branches of the rather shallow canopy layer are at the top of a tall straight trunk. The trunk elongates rapidly enabling the tree to get and keep its place in the sun. Thus the internodes between the branches are long. The lower shaded branches only persist for a relatively few years, then die, and fall off. Once the limb is lost the knot is covered producing a clean trunk and "clear lumber". This infers there were few limbs to start with on the ancient trees, but there were some. Limbs and spreading roots on dead trees are lost from the main trunk rapidly, the smaller they are, the faster they go. Decomposers, microorganisms to insects, are efficient recyclers. Organic material is thus decomposed, chewed up, ground up, hydrolyzed, or oxidized - gone in no time. The forest floor was continually cleared of litter before becoming mineralized. What small full-round specimens that were petrified have been picked up for over a 100 years. Collectors over the years have cherished and coveted small limbs, thus small limbs are precious.

The large volumes of airborne rhyolitic materials of the Triassic were initially carried only as far as the energy of the explosion transported them, or the force of the wind blew them. From localized to far flung places, some if not all of the materials of an eruptive event or episode were picked up by water to begin a transient journey ultimately contributing to the build up of the Chinle deposits. Tropical storms, particularly heavy monsoonal storms, washed ash from leaves and the ground's surface, and deliver it to streams. These flowing waters of tributary drainages and whole drainage basins concentrate the dispersed fallen ash, especially when a large drainage area contributed its dispersed mantle to a restricted region. *Woodworthia* may have filled the niche now occupied by the Bald Cypress of today's swamps, even producing knees. More species grow on the higher hammocks; and on the levies and banks as riparian communities. The initial uplifts and volcanos to the west and south, with rejuvenated uplifts in the eastern highlands provided the coarse sediments to the Shinarump Conglomerate, the lowest Chinle member. These sediments filled the channels and valleys whose irregular surface topography represents the unconformity developed during the erosive mid Triassic. Logs and other organic debris came to rest in and along these aggrading river channels.

8 Agate Bridge
Petrified Forest National Park, Arizona

9 "Rainbow Wood"
Dobell/Patton/Gray digging area
west of Petrified Forest, Arizona
46.5 cm diameter, photo: Foster

10 *Araucarioxylon arizonicum*
Milky Ranch, Arizona, Thompson collection, 40 cm diameter, photo: Foster

11 *Araucarioxylon arizonicum*
Sanders, Arizona, 44 cm diameter, photo: Foster

During the lower Petrified Forest/Blue Mesa time, the earlier/older part of the Chinle, the dominant trees were *Araucarioxylon arizonicum*. This species is now memorialized by the numerous long logs and large log sections in the southern part of the park. The log forming Agate Bridge in the Sonsela Sandstone is also *Araucarioxylon* (Fig. 8). From superficial observation, one may think that *Araucarioxylon arizonicum* makes up 90% of the logs in the Chinle, particularly in the lower part of the formation. But looking at stem cross sections of different wood specimens with a 20x hand lens reveals that there are at least 3 and maybe 4 different species all passing for *A. arizonicum*. Until a more detailed study is made of the cell and stem structure of these other types, with comparisons to the type *A. arizonicum* specimens described by Knowlton in 1888, referencing two logs collected in 1879, one from Lithodendron Wash, a site within the present National Park, and one from north of Fort Wingate, New Mexico - only *A. arizonicum* is recognized. Presented photos depict some of the variation in the petrified wood found in the Chinle. Mineralization expressed in color, color pattern, and cell preservation is characteristic of particular localities. Some identifiable plant taxa are very localized, others more widespread. Species composition is characteristic of certain localities, which is more or less an expression of the ancient habitat. Within many areas all of the logs, in external appearance, are similar. Without close scrutiny, species that are different pass as the same thing - "Arizona wood".

Araucarioxylon arizonicum

Araucarioxylon has moderate size tracheids that are relatively uniform in tangential diameter. It has indistinct

or no definition of growth rings. This species, the dominant tree of the Chinle forest (Figs. 10-11), was up to 1.5 m dbh, and may have reached 60 m tall. The type specimens are at the Smithsonian, Washington D.C.
As Laurasia moved northwesterly away from the equator on a tectonic trip for some 15 to 20 million years, there was a resulting northward latitudinal change of perhaps 2.25 to 3 degrees representing 250-325 km. As a consequence of this drift there was a gradual climatic change, going from a humid tropical climate with little seasonal variation in precipitation, to a climate, still tropical, but with a seasonal drier monsoonal period. This seasonal climate is reflected in growth rings, particularly during Black Forest time. As time progressed, the diversity of the vegetation increased as expressed by the petrified wood types found today in the later/upper parts of the Chinle, i.e. trees of *Woodworthia, Schilderia* and "*Ginkgo*".

Woodworthia arizonicum

A living *Woodworthia* of the Triassic would have looked rather strange in the sense of today's conifers. It had short thin branches/short shoots with clusters of leaves decorating its trunk and limbs. Today, there is a pine species, *Pinus rigida*, growing along the St. Lawrence River, Quebec, and south through the eastern U.S., that has these pompons. *Woodworthia* in stem cross section shows prominent short-shoot traces connecting the pith to the short-shoot scars/knots on the outside of the stem. These short-shoot traces are a section through the pith with only a very thin sheath of surrounding xylem. *Woodworthia arizonicum* (Fig 13) has narrow, 2.5 to 3 mm, short-shoot traces leading to dimpled/projecting short-shoot knots.

12 *Woodworthia* sp.
Egg Canyon, Circle Cliffs, Utah
43 cm diameter

13 *Woodworthia arizonicum*
north of Winslow, Arizona
41.5 cm diameter, photo: Foster

14 „*Dadoxylon*" sp.
Down toward the Colorado River, Utah
Lyman collection, original size

Woodworthia sp.
Another *Woodworthia* (Fig. 12), as yet undescribed, with 5 to 8 mm wide short-shoot traces is showing up in Holbrook and St. Johns, Arizona.
I acquired a specimen of this type from Egg Canyon, Utah in the 1970's, and saw one from the Kanab, Utah area in 1996.
Still another, undescribed, *Woodworthia* from the Long H Ranch, Arizona and west of Blanding, Utah has short-shoot traces that are only about 1 mm wide.
The use of the genus name *Araucarioxylon* has been restricted to designation of trees of the Mesozoic. *Dadoxylon*, in general, is used for trees of the Palaeozoic period. Both genera have similar araucarioid cellular characteristics within the xylem. But, there is a distinct difference in the stem structure between the two genera. *Dadoxylon* has a variable sized but large pith, while *Araucarioxylon* has a small, visually non discernable, pith. This contrasting pith/stem difference is the distinguishing feature, not whether the tree was from a Palaeozoic versus a Mesozoic time interval. The generic names are not synonymous. Daugherty recognized this in his naming of *Dadoxylon chaneyi*, a stem with a large pith, from the Chinle, Triassic-Mesozoic. Stems from the Permian of Chemnitz, Germany, lacking a large pith, should more properly be referred to as *Araucarioxylon* rather than *Dadoxylon*. In Europe, the distinctive large pith of *Dadoxylon* may be referred to as the *Artisia* mark.

Dadoxylon chaneyi
In *Dadoxylon chaneyi* the pith is 16 to 25 mm in diameter, with peripheral small distinct serrations formed by the primary xylem centers of growth. This species has small tracheids - 15 µm in tangential diameter. The type material is in the collections of the Museum of Palaeontology, University of California, Berkeley.

"*Dadoxylon*" sp.
Another "*Dadoxylon*", but probably a new genus and certainly a new species, has a very large star patterned pith, the tips of the rays splaying to 7 cm, with a central core to the pith of 2-3 cm. In squashed specimens the pith star forms a distinct coarse zigzag pattern (Fig. 14).

"*Ginkgo*" sp.
The *Ginkgo biloba* of today is the last remnant of a group of gymnosperms that have a long and diverse heritage. Considering the wide ranging distribution of horticultural plantings, is hard to figure why the group dwindled to its relict status. Leaves of *Ginkgo* are relatively common in the fossil record, but petrified wood is assumed to be a rare commodity (Fig. 15). This assumption may not be well founded. Superficial observation will not differentiate a *Ginkgo* from an araucaria. *Ginkgo* has large parenchyma cells scattered among the tracheids. The tracheids are small, but exhibit a wide range of diameters within and between adjacent radial files of the xylem cells. The living *Ginkgo* is susceptible to a heart rot fungus. Fossil wood preserves the same sort of problem.

Undetermined stem
Another specimen, of a new taxon, has surfaced in Holbrook and has just been shown to me. It has a large pith, about 1 cm in diameter, similar in cell structure to, but smaller than *Dadoxylon chaneyi*. This specimen has vertical and horizontal resin canals - resin canals are not present in *D. chaneyi*. With these features this specimen is not going to be considered *Dadoxylon* or araucarian - it is more like *Pinus*, but with a "large" pith.

Schilderia adamanica
The distinctive species, *Schilderia adamanica* (Fig. 16), was a large tree. Setting *Schilderia* apart from other petrified woods are the wide rays,

15 „*Ginkgo*" wood
Sperlock Ranch, southern Navajo, Arizona
15.5 cm diameter

16 *Schilderia adamanica*
Red Canyon, Utah, photo: Moore

clearly visible to the naked eye. With magnification, the chevron patterned cell arrangement in the rays of *Schilderia* can be seen. This stem and ray anatomy is mirrored in the living, shrubby Ephedra, Mormon Tea. These characters seem to tie *Schilderia* to the living Ephedraceae. *S. adamanica* is a good candidate for the species shedding the ephedracious pollen found in the Chinle. *Schilderia* appears to have been very susceptible to a heart rot fungus which produced hollow tree trunks. Thus, many specimens are seen that were crushed prior to mineralization.

Lyssoxylon grigsbyi
Three fragmented specimens of this species have been collected. The first came from an island in the Rio Puerco west of the National Park and SE of Adamana; two others came from a few miles east of Fort Wingate, New Mexico. This species was first thought of as cycadeoid, then redefined as a cycad. Its stem was approximately 8 cm in diameter. Its height is hard to determine when based on only a 13 cm long fragment.

Charmorgia dijolii
Three specimens of this species have been found. The type specimen of this species was found in the Joseph City area. The type locality for this specimen has been variously given as the lower Petrified Forest member a few miles northeast of Joseph City, and/or near Tanner Crossing on the Little Colorado River, and/or about 7 km north of the Cholla Electric Generating Plant - three different spots. This last reference is the accurate site. This species was originally referred to as a cycadeoid, but when finally described was placed with the cycads. The stem is about 25 cm in diameter, but relatively short, something over 10 cm tall. The type specimen is on display at the Museum of Northern Arizona. The two other specimens, paratypes, were found in the southern part of the National Park. One is relatively complete, although dorsoventrally flattened. The other specimen was nearly complete when found, but most of it has been lost and is now represented only by a few fragments. The paratype specimens are in the National Park collections.

Itopsidema vancleavei
This plant is known from 11 fragments of one stem. The type locality is within the National Park, north of Agate Bridge, in a bentonitic shale, below the Newspaper Rock Sandstone layer. The plant was more than one meter tall. Its stem was only about 1.3 cm in diameter, but the mantle of roots and petioles produced a pseudotrunk some 10 cm in diameter. The segments of the type specimen are in the palaeobotanical collection of the Petrified Forest National Park.

Donwelliacaulis chlouberi
Another fern represented by several hundred pounds of material in more than 200 fragments, attributable to several dozen individual specimens, was collected near Winslow, Arizona. It came from the lower Petrified Forest member. The largest specimen is about 30 cm in diameter and more than a meter tall, but this may be less than half the height of the trunk. This fern material is at the Museum of Northern Arizona.

Neocalamites cf. *virginiensis*
Neocalamites fossils in-the-round are found at a number of localities, Blue Forest in the National Park, on slopes of Poleo Mesa near Coyote, New Mexico; on the western slope of Zuni Mountain between Zuni and the Little Colorado River; from the old copper mines near Abiquiu, New Mexico, and from north of St Johns, Arizona. The fossils are sandstone pith casts from the stems and rhizome casts. The pith casts may be 20 cm in diameter and over 1 m in length. The rhizomes may be 6 cm in diameter.

Special Features
Some petrified wood collectors want "special features" in the specimens

they obtain. If one looks, there are a wide range of fungus and bacterial diseases, insect infestations, and fire scars present in fossil wood. Trees of the fossil forest senesed and died, as trees do today. Standing dead trees become snags and eventually fall, joining their prostrate contemporaries. Virtually all of the logs show signs of disease, many are hollow. The cell structure of many logs is poorly preserved, indicating the general process of decomposition after death. These "imperfections" only hint at stories of life and disease in the long past ecosystem.

Polyporites wardii

The pockets in this pocket rot fungus tend to be radially oriented. They are elongate ovoid 2.5-8.5 cm long x 1 to 3.5 cm wide, in radial aspect; and oval, 1-2 cm thick, in cross section (Figs. 17-18).

One more, still unnamed Pocket Rot Fungus (Fig. 20) shows pockets that are much smaller than in *P. wardii*. They are radially oriented, ovoid 15-20 mm long, 7-8 mm wide, in radial aspect; and lenticular, 4 mm thick in cross section.

Lightning Scars

Lighting on striking a treetop may blow that top to pieces. The bark and some wood can also be blown off as the charge travels down the side of the trunk to the ground. The trunk is not burned and over time the scar is overgrown (Fig. 19).

Insect Damage

A sensitive observer can find evidence of insect activity, fine bark beetle engraving to large flathead borer burrows, in the fossil woods.

17 *Polysporites wardii*
Large pocket rot fungus
Patton Ranch, Arizona, Rose collection
71.5 cm diameter, photo: Foster

18 *Polysporites wardii*
Large pocket rot fungus
Patton Ranch, Arizona
height 9 cm

19 *Araucarioxylon*
Patton Ranch, Arizona
30 cm diameter

20 Still undescribed and unnamed pocket rot fungus
Patton Ranch, Arizona, height 49 cm

WILLIAM D. TIDWELL

Die Osmundaceen - eine uralte Farnfamilie

The Osmundaceae - a Very Ancient Group of Ferns

The Osmundaceae is one of the oldest known living fern families in both the Northern and Southern Hemispheres. It's earliest definite fossil records are from the Permian period. Throughout the Mesozoic era, leaf compressions and petrified stem material indicate the family was geographically widespread. During the Tertiary, however, it appears to have decreased in numbers and distribution. Those family members that have survived to the present may, in a sense, be referred to as "living fossils."
The Osmundaceae is best known for its three living genera, *Osmunda*, *Todea*, and *Leptopteris*. *Osmunda* is cosmopolitan in distribution and consists of an erect (Fig. 1) to creeping, underground, nearly equal branching stem. The apex of each branch bears a thick mantle of leaves, some of which are over two meters long. It also produces a very large number of persistent, wiry roots from its stem. Stems, leaf bases, and roots of *Osmunda* may become very massive, forming large hummocks that can be more than a century old.

Todea and *Leptopteris* (Fig. 2) are restricted to areas of the Southern Hemisphere. *Leptopteris* is confined to Australasia and the South Sea Islands, whereas *Todea*, represented by one species, grows in South Africa and Australasia regions. They are small tree ferns with leaves forming a crown near their tops. Eighteen living species are included in these three genera, and there are 150 extinct osmundaceous species worldwide.

Most of these are based on compression-impression leaf material with about 50 species described from petrified stems usually with attached leaf bases and roots. Presently, these fossil species are included in approximately 18 fossil genera. Those genera similar to the fossil *Thamnopteris* represent an extinct group in the Osmundaceae that does not directly relate to the other members of the family. Accordingly, they are placed in a separate subfamily, the Thamnopteroidae. The remainder of the family, both living and fossil, is then assigned to the subfamily, Osmundoidae.
Anatomically, the stem of members of the Osmundoidae is very distinctive. This stem structure is very similar in most living and fossil genera of this subfamily. The bulk of the stem consists of cortices that are sometimes indistinguishable from the mass of leaf bases. In general, the C-shaped leaf traces are surrounded by a sclerenchymatous sheath known as the sclerotic ring (Fig. 26). The vascular tissue (stele) of the stem is somewhat dictyostelic in general with relatively long narrow leaf gaps existing in the xylem cylinder (Figs. 2-3, 13). In cross section, the water conducting xylem cylinder shows a ring of xylem strands around a central pith, outside of which there is the food conducting phloem and a continuous endodermis (Figs. 3, 8). The xylem strands generally appear somewhat elongated in a radial

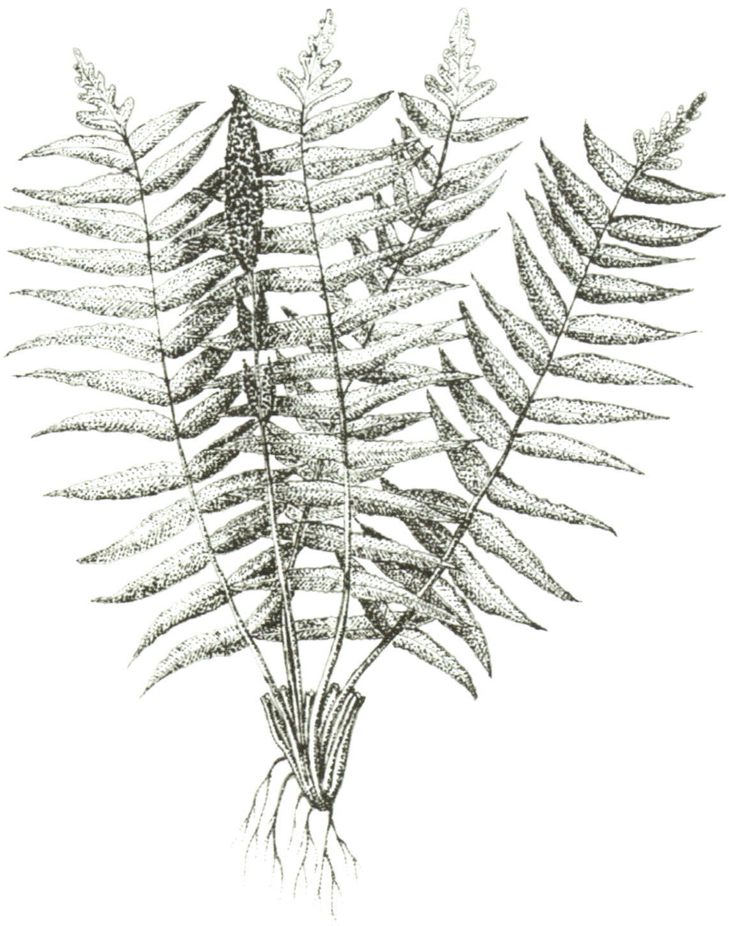

1 Living *Osmunda*

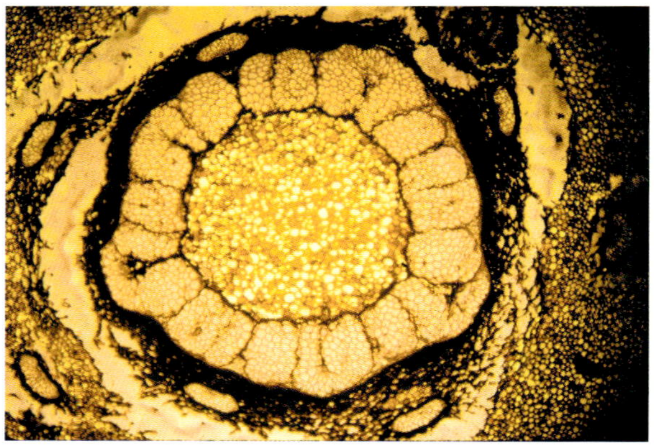

2 Stem of living *Leptopteris* (xs), 20 x

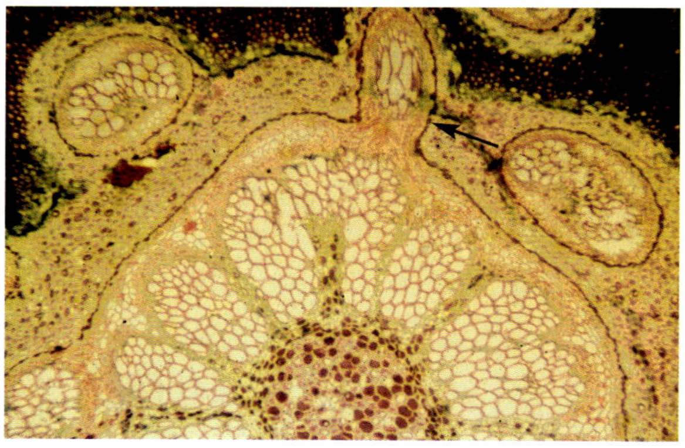

3 Portion of a stem of living *Osmunda* (xs)
Note root departure (arrow), 30 x

4 *Palaeosmunda williamsii* (xs)
Queensland, Australia, 0.5 x

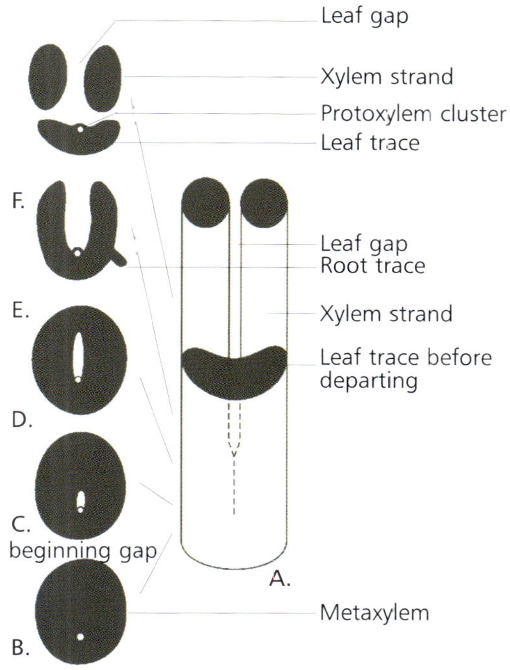

5 Leaf trace and gap formation at different levels of a xylem strand; protoxylem (white dots) (modified from Miller 1967)

direction (oval) and occur singly or in close pairs (Fig. 5A). In the living genera and most of the fossil, the strands can be oval, O, U, or S-shaped, depending on their position within the xylem cylinder during leaf trace development or departure. They are oval before gap formation begins (Fig. 5B), O-shaped as the gap and trace start to form (Fig. 5C, D), U-shaped before leaf trace departure (Fig. 5E), two radially aligned xylem strands remain after trace departure (Fig. 5F), S-shaped as adjacent strands fuse closing the gap, thus returning to oval and eventually beginning the trace formation again. Root traces originate from the leaf traces, usually before the U-shaped leaf trace departs (Figs. 3, 5E, 8).

Petrified Axes

Thamnopteroidae

Stems of the genera of the Thamnopteroidae, are surrounded by a mantle of compact petioles and a small number of roots. For example, the arborescent stems of *Thamnopteris* and *Zalesskya*, both from the Upper Permian of Russia, generally consist of a solid cylinder of xylem (protostele), surrounded by a relatively thick cortex and wider zone of leaf bases. The xylem cylinder of the stems is divided distinctly into a central zone of short tracheids that probably functioned for water storage, and an outer zone of longer tracheids for water conduction. The cylinder is surrounded by a continuous layer of phloem, a feature that links these genera with the Osmundaceae. The spirally arranged leaves originate from the outer edges of the xylem cylinder without forming leaf gaps. The traces occur first as oval strands that eventually narrow to a horseshoe-shape at higher levels of the petioles. Stipular wings containing 1-3 clusters of thick-walled cells are also present in *Thamnopteris*.

Osmundoidae

Palaeosmunda is a primitive member of the Osmundoidae from the Permian of Queensland, Australia. It has a stem characteristic of the living

6 *Ashicaulis wadei* (xs)
Morrison Formation near
Moab, Utah, original size

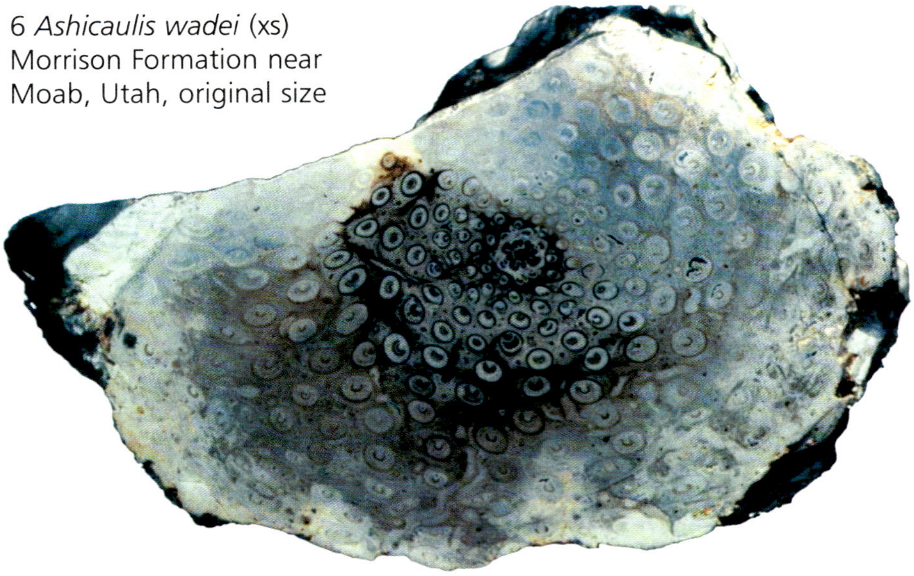

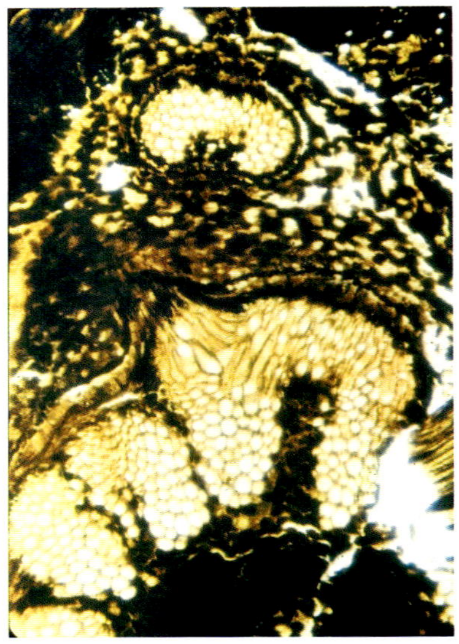

7 *Ashicaulis wadei* (xs)
Trace development
note departing root, 22 x

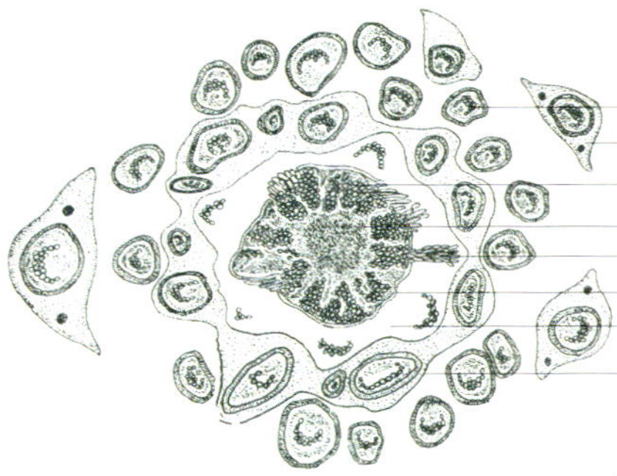

8 *Ashicaulis wadei*
trunk in transverse section, 2 x

osmundas (Fig. 4), having a well-developed pith, and leaf gaps forming in association with C-shaped leaf traces in its xylem cylinder. The sclerotic rings of the petioles are rhomboidal in shape and become laterally extended at higher levels of the petioles.

The history of naming the petrified axes (stems) assigned to the Osmundoidae from the Mesozoic is rather involved. The form genus *Osmundites*, used for a long period of time, is invalid since it was used previously for misidentified cycadophyte foliage. After *Osmundites* was declared invalid, the new genus, *Osmundacaulis*, was then proposed for these fossil osmundaceous axes. At that time, the known species were placed in three informal groups: the "*O. brazilensis*" group, the "*O. herbstii*" group, and the "*O. skidegatensis*" group. The species of the "*O. brazilensis*" group were eventually separated out to form the basis for the genus *Guairea* and, thus, the family Guaireaceae (Tidwell, in this volume). Members of the "*O. herbstii*" group were divided between *Millerocaulis* and *Ashicaulis*, and the "*O. skidegatensis*" group remained as the only members of *Osmundacaulis*.

Millerocaulis and *Ashicaulis*

Millerocaulis and *Ashicaulis* are rhizomatous and occasionally arborescent with small, often branching stems, surrounded by a mantle of leaf bases and roots (Fig. 8). The stems contain a sclerotic (fibrous) outer cortex usually much wider than their parenchymatous or sometimes parenchymatous and sclerenchymatous inner cortices. In occasional species, the two cortices are of nearly equal width. They both have C-shaped leaf traces originating from their xylem cylinders. Their leaf bases are stipular

9 Little Swanport locality near Swansea
Tasmania, Australia

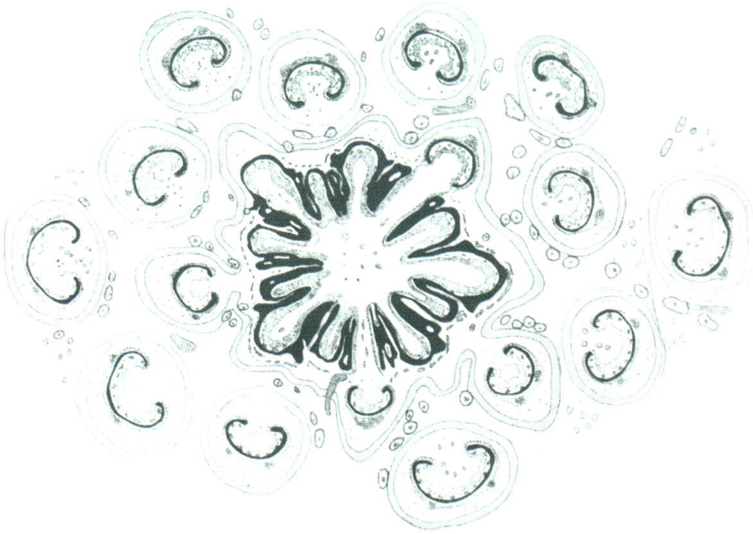

10 Stem and adjacent petiole traces of
Osmundacaulis nerii (xs)
Xylem (black), sclerenchyma (light stipple), outer cortex and outer sclerotic ring or trace (darker stipple), externally to the xylem, the dashed lines delimit the endodermis and phloem and internally, these lines indicate the xylem sheath; roots occur throughout the section
Tasmania, Australia, 2.5 x

11 *Millerocaulis* sp. (xs)
Note light colored xylem cylinder
without leaf gaps (arrow)
Queensland, Australia, original size

and adventious roots arise singly or in pairs from the traces.

Millerocaulis and *Ashicaulis* appear to be very closely related genera with most species of both genera occurring in the Southern Hemisphere, notably Australia (Figs. 6, 11). The basic difference between the two is whether numerous well-defined leaf gaps that form like most of the Osmundoidae are present (*Ashicaulis*, Figs. 7-8) or whether the stem lacks leaf gaps or has occasional poorly defined gaps (*Millerocaulis*). In *Millerocaulis*, the leaf trace originates with the insertion of parenchyma into the xylem cylinder. This insertion, or pocket that forms, continues to expand towards the pith at the same time that the portion of the xylem cylinder, which eventually becomes the leaf trace, expands outwardly into the inner cortex. The insertion or pocket does not normally reach the pith but extends one fourth to three fourths distance to the pith before the U-shaped trace departs, forming an incomplete gap. After the trace leaves, the upper parts of the cylinder ultimately unite, thereby completing the cycle. The parenchyma sometimes extends to the pith, forming the occasional gap. A root trace diverges from the leaf trace before the departure of the latter.

Millerocaulis species have been described from "Cattle Downs" northwest of Wandoan and the "Limewood" Station northeast of Orallo, Queensland. They appear to have come from the Middle Jurassic Birkhead Formation, whose depositional environment was very likely swampy. *Osmundacaulis hoskingii* was recorded from nearby "Mt. Organ" Station and *Grammatocaulis donponii* from "Warrawoona" Station southeast of "Cattle Downs". The small osmundaceous stem, *Millerocaulis embreei*, was described from marine sediments of the Early Cretaceous Upper Chickabally

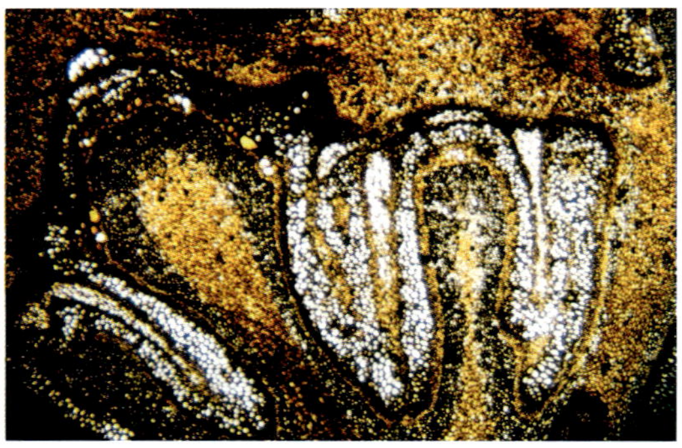

12 *Osmundacaulis nerii* (xs)
Micrograph of leaf trace formation, 10 x

13 *Osmundacaulis andrewii* (xs)
Micrograph illustrating the numerous leaf gaps of this species, Tasmania, 10 x

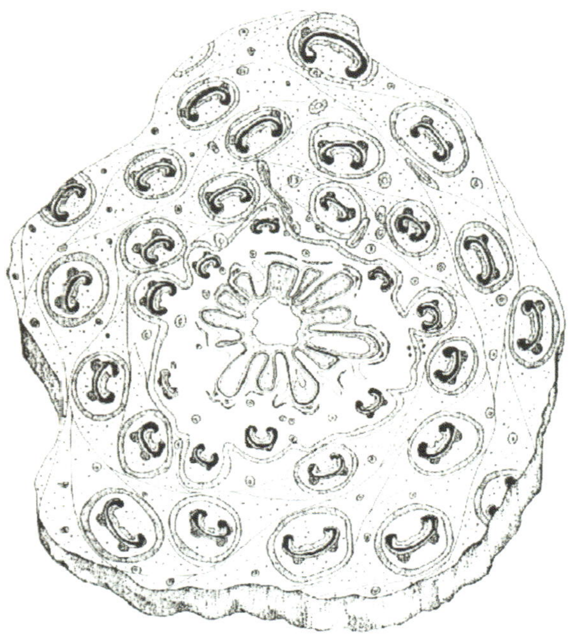

14 *Osmundacaulis lemonii*, (xs) from near Castle Dale, Utah, U.S.A.
Note that it is very much like *O. nerii*, 2 x

15 *Osmundacaulis griggsii* (xs) from Lune River locality, Tasmania, Australia 0.5 x

Member of the Budden Canyon Formation near Ono, California. This species is the youngest known of this genus and the first to be reported from the Northern Hemisphere.

Ashicaulis species occur at several sites throughout the world. Six species were collected from five localities in Tasmania. One near Ida Bay in southern Tasmania, and four from eastern Tasmania near Swansea (Fig. 9). Those from the Swansea area were collected from gravels, but appear to have been reworked from the nearby Upper Triassic coal measure sequence.
Ashicaulis wadeii was reported from the Upper Jurassic Morrison Formation near Moab, Utah, U.S.A. Specimens of this species have also been collected from a site east of Ferron, Utah associated with *Osmundacaulis lemonii*. Some specimens of *Ashicaulis* were found in the Queensland, Australia affiliated with specimens of *Millerocaulis*. *Ashicaulis* species also occur in the Jurassic of India and New Zealand.

Osmundacaulis
Osmundacaulis (sensu stricto) is generally arborescent with small to usually large stems, upwards of 45 cm in width with steles that are dictyostelic, often with phloem inside and outside of the xylem (amphiphloic; Figs. 10, 12, 14-16). The xylem cylinder in this genus can be wide, as much as 25 or more cells (tracheids) thick and divided into a relatively high number of xylem strands (Figs. 12-13). The inner cortices of the *Osmundacaulis* species are wider than their outer, rarely nearly equal in width. Leaf traces are strongly curved, but often become flattened outwardly. The petioles generally contain continuous or lobed sclerenchyma lining the concavities of the C-shaped xylem strands and sclerenchyma that is often scattered in their cortices. Leaves were coiled (circinate) when they were young (Fig. 17).

Presently, fifteen species of *Osmundacaulis* have been described from strata of mid-Mesozoic age throughout the

16 *Osmundacaulis* sp. from Lune River locality, Tasmania
Dernbach collection, original size

17 *Osmundacaulis janii* circinate leaves
Lune River locality
Tasmania, Australia
nearly original size

18 Lune River locality, Tasmania, Australia

world with thirteen of these species being from the Southern Hemisphere and only two from the Northern. The Lower Cretaceous *Osmundacaulis* skidegatensis from Alliford Bay, Skidegate Inlet, Queen Charlotte Islands, British Columbia, Canada was the first dictyostelic osmundaceous species recorded. Later *O. natalensis* and *O. atherstonei* were reported from South Africa. *Osmundacaulis natalensis* is assumed to have come from the Lower Cretaceous Ndbana beds, Mzinene River near Hluhluwe (Zululand), whereas both *O. atherstonei* and *Ashicaulis kolbei* were reported from the Lower Cretaceous Uitenhage Series near Geelhoutboon in the Sundays River Valley (*O. atherstonei*) and Herbertsdale (*A. kolbei*), Cape Province. Two unpublished species of *Osmundacaulis*, *O. stumkei* and *O. zululandensis*, were also identified from the Lower Cretaceous Mzineni Formation of South Africa. *Osmundacaulis stumkei* was collected from Farm Vergenoeg near the Mzineni River and *O. zululandensis* is from the M'hlosinga district just north of the road to Bumbeni as it crosses Manuan Creek, Natal. *Osmundacaulis* sp. (Figs. 19, 19a) from the province of Rio Negro, Argentina appears to be the first *Osmundacaulis* reported from South America. *Osmundacaulis hoskingsii* with two varieties, var. *hoskingsii* and var. *tabulatus*, was described from the Jurassic of Queensland, Australia, and *O. lemonii* (Fig. 14) was collected from the Upper Jurassic Morrison Formation near Ferron and Moab, Utah. *Osmundacaulis nerii* (Fig. 10) and *O. jonesii* occurred near Lune River, Tasmania, Australia, where *Osmundacaulis* specimens are often referred to as "manfern" and the petiole traces as "eyes." Six species were further described from the Lune River area. These include *O. janii, O. richmondii, O. pruchnickii, O. griggsii, O. andrewii*, and *O. tasmanese*.

The Lune River locality (Fig. 18) is south of Ida Bay in southern Tasmania. Plant remains from this locality other than *Osmundacaulis* species are the tree ferns, *Cibotium tasmanese* and *Oguracaulis banksii* (Tidwell, in this volume), the leaf forms *Conopteris websterii*, *Cladophlebis indica*, *Pachypteris* sp. cf. *indica*, *Otozamites* sp., ? *Pterophyllum* sp., and *Conites* sp. The specimen of *Otozamites* sp. was partially petrified and, thus, gave an indication of its internal tissues. The fossil material appears to have been derived from a basalt flow that occurs on a ridge above and west of the collection site. This was demonstrated by a specimen of *O. nerii* which was collected with some of the basaltic remains still attached. The basalt, originally considered Tertiary, is now known to be mid-Mesozoic, probably Early Cretaceous in age.

Aurealcaulis

Aurealcaulis is unique in that its leaf traces form by fusion of segments of adjacent xylem strands either in the inner or outer cortices, or outside of the stem, depending on the species . The gaps generally remain open or only partially close (Figs. 21, 23, 24), which is the opposite situation in *Millerocaulis* where the gaps never or only occasionally open. Root traces usually originate from the developing leaf traces in the inner cortex rather from the traces before they diverge from the xylem cylinder as in the other osmundaceous genera. Presently, *Aurealcaulis* is known only from the U.S.A. It has been collected from the the Eocene of New Mexico (2 spp., Figs. 22, 24, 25, 26),

19 *Osmundacaulis* sp. (xs). First *Osmundacaulis* to be reported from South America Rio Negro, Argentina, original size

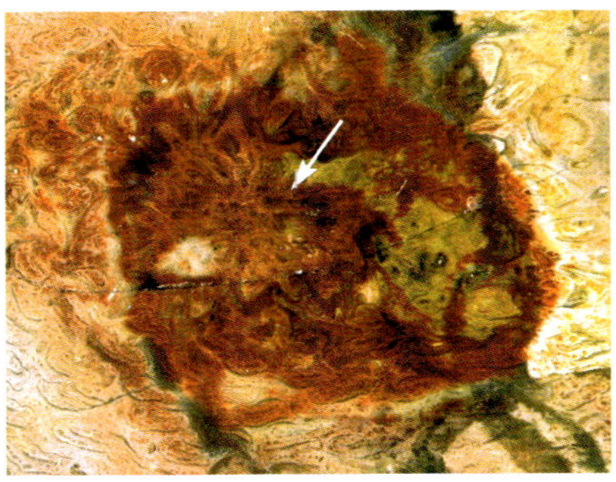

19a *Osmundacaulis* sp.
Detail showing the stem and leaf trace formation (arrow), 15 x

Palaeocene of Wyoming (1 spp.), and the Lower Cretaceous of South Dakota (2 spp.) and Nebraska (1 spp.).

Aurealcaulis crossii, the first species of this genus to be described, was collected from the Palaeocene Fort Union Formation southeast of Rock Springs, Wyoming (Fig. 20). A few specimens were found in place at this site in the thin-bedded, fluvial carbonaceous sediments from which they were removed. Compressions of fern pinnules assigned to *Allantodiopsis erosa*, conifer needles of *Glyptostrobus* sp., various dicot leaves, and *Osmunda greenlandica* (Fig. 27) were also found associated with specimens of *A. crossii*.
The three species, *A. burgii*, *A. dakotensis*, and *A. nebraskensis* occurred as water-worn cobbles in a gravel of the Eocene Chamberlain Pass Formation in South Dakota. and Nebraska. They were associated with various species of the fern *Tempskya*, the tree fern, *Nishidacaulis burgii*, specimens of *Cycadeoidea*, and the leaf compressions, *Gleichenia comptoniaefolia* and *Cladophlebis* sp., al Early Cretaceous in age. They appear to have been eroded from the Lakota Formation of the Black Hills of western South Dakota.

20 *Aurealcaulis crossii* collecting site near Rock Springs, Wyoming, U.S.A. Note pile of fossils at lower left of photo.

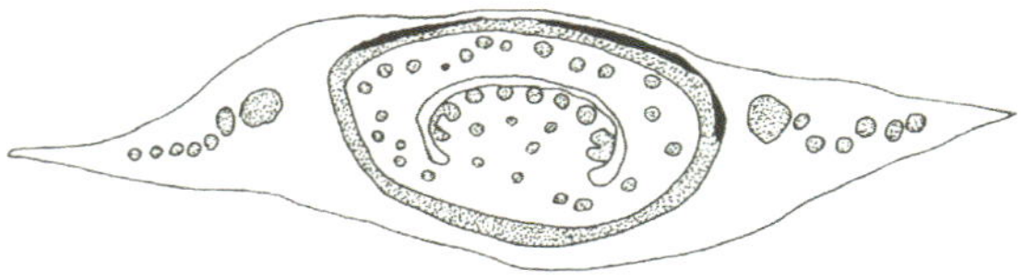

22 Petiole of *Aurealcaulis bransonii* showing a C-shaped leaf trace surrounded by a sclerotic ring and stipular wings transverse section, 5 x

21 *Aurealcaulis crossii* 0.5 x

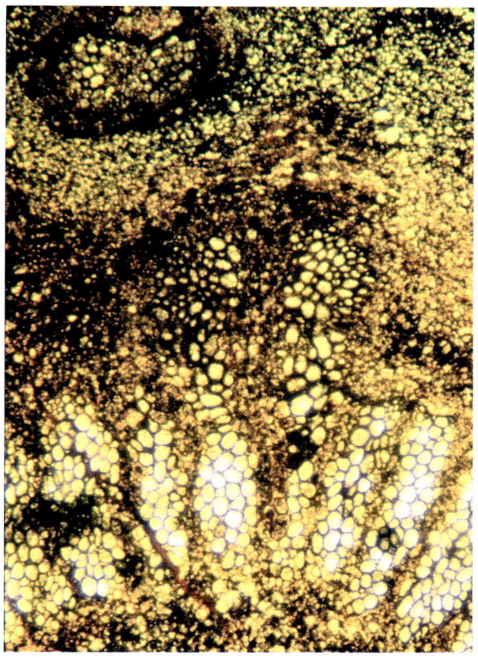

23 Detail of xylem and leaf trace development of *Aurealcaulis crossii* Note extension of the xylem strands before fusion, cross-section, 25 x

24 Stem of *Aurealcaulis moorei* transverse section, 5 x

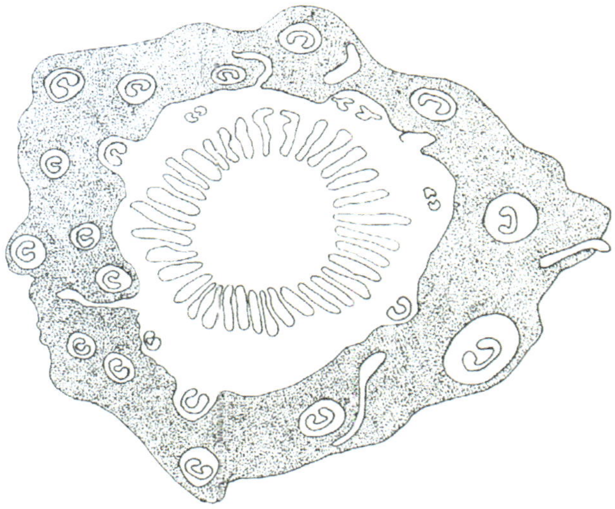

Osmunda

Osmunda has several fossil species, as well as living. They occur at various localities around the world, but only a few will be considered here. Petrified underground stems (rhizomes), for instance of *Osmunda pluma* and *O. arnoldii*, occur in boulders of different sizes that are scattered over wheat fields approximately five miles west of Glen Ullin, North Dakota, U.S.A. The origin of these boulders were traced to two or more thin beds in the Palaeocene Fort Union Formation. Many of the fern rhizomes were preserved between two or three pieces of flat, fossil conifer wood (Fig. 29). The longitudinal axes of the fossil wood usually parallel the longitudinal axes of the well preserved rhizomes. The fern rhizomes appear to have been directly related to the wood when they were living and were more or less preserved in place.

The first species of petrified rhizomes to be directly assigned to *Osmunda* was *O. dowkerii*, which was first reported from Eocene strata of Herne Bay, United Kingdom. Specimens from the Eocene Clarno Formation in Oregon, U.S.A. described initially as *Osmundites chandlerii* were subsequently placed with *O. dowkerii* by Miller (Fig. 28). He also noted three other specimens of this species. One was from a boulder found on the beach of the Oregon coast, another from Utah, and a third from the Black Hills of South Dakota. *Osmunda* oregonensis was also collected in the Clarno Formation from the Crooked River Valley east of Post, Oregon on Lost Creek. The strata at the collection site is composed mostly of lava flows with intervening beds of ash and tuff. *Dennstaedtiopsis, Acrostichum*, and *Equisetum* are other petrified fern or fern allies reported from this formation in this general area.
A variety of fossil wood, leaves, and seeds also occur in this formation. The occurrence of *O. cinnamomea* was established in the Miocene-Pliocene and *O. wehrii* in Late Miocene by fossil rhizomes from near

25 Actual specimen of *Aurealcaulis moorei*
Near Farmington, New Mexico, 4 x

26 Detail of *Aurealcaulis moorei*
petioles with stipular wings
transverse section,10 x

Yakima, Washington, U.S.A. The rhizomes of *O. wehrii* occurred near the contact of the Roza Basalt and the overlying Priest Rapids Basalt. *Osmunda iliaensis* was collected at or near the contact of the Miocene and Pliocene at Ilia, near Schemnitz, Hungary, and *O. nathorstii* was reported from Spitzbergen.

The petrified axes of the Osmundaceae appear to be universally very beautiful and, scientifically, very interesting material. Much has been learned about members of the family over the past hundred and fifty years and, undoubtedly, more awaits to be discovered about them in the future.

27 *Osmunda greenlandica* collected next to *A. crossii*, 2 x

28 *Osmunda dowkerii* (xs) Clarno Formation, Oregon original size

29 *Osmunda pluma* (xs, arrow) within fossil spruce wood Fort Union Formation, North Dakota, 2.5 x

WILLIAM D. TIDWELL

Die Guaireaceen - eine ausgestorbene Farnfamilie

The Guaireaceae - an Extinct Fern Family

Guaireaceae, an extinct family, was proposed for fossil tree fern stems similar to those of the Osmundaceae but having different or lacking certain characters that occur in the latter family. These characters include a cortex in Guaireaceae that is not divided into two layers (inner and outer cortices) and having leaf (petiole) bases that lack both sclerotic rings around the petiolar vascular traces and stipular wings, features that occur in the Osmundaceae.
The outcurved tips of the petiole traces of *Lunea* and *Itopsidema* may also represent a diagnostic character for the Guaireaceae.
Guaireaceae was originally a monogeneric family. Two species previously placed in the Osmundaceae, *Osmundacaulis brazilensis* from the Permian or Jurassic of Brazil and *O. carnieri* from the Permian of Paraguay, were combined together and assigned to the new genus *Guairea* as *G. carnieri* (Fig. 1). Another arborescent species, *G. millerii*, was later described from the Upper Permian of Paraguay, and based on new stratigraphic evidence, both species were placed in the "Permian Osmundalean Complex". Later the family was emended to accommodate another form, *Lunea jonesii*, from the Lune River locality of Tasmania. *Itopsidema vancleavei* from the Upper Triassic Chinle Formation in the Petrified Forest National Park, Arizona and *Donwelliacaulis chlouberii* (Fig. 4) from the Middle Triassic Moenkopi Formation, also in Arizona, were eventually included in the family as well.

1 *Guairea carnieri* (xs)
showing stem and leaf traces, Paraguay
Dernbach collection
17.5 cm diameter

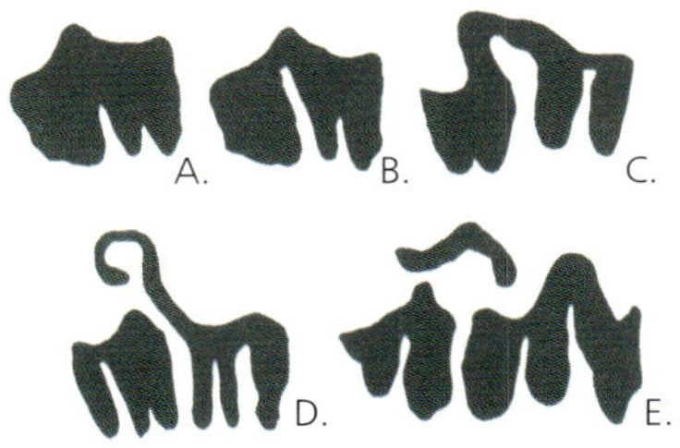

2 *Guairea millerii* (xs)
Formation of leaf trace
A, B. Begining of leaf gap on pith side of xylem cylinder (black)
C. Trace widening outwardly on cortex side
D, E. Leaf trace departing, first on one side (D) and then the other (E)
modified from Herbst 1981

3 *Lunea jonesii* (xs)
Tasmania, Australia, 0.8 x

Guairea

Guairea has a very large stem, parenchymatous pith and cortex, and leaf gaps that form by a process of "invagination" where the xylem strands split apart beginning on the pith side of the xylem cylinder and expand outwardly (Figs. 1, 2). Leaf traces are shallowly curved, to C-, V-, or Y-shaped (Fig. 1), some with islands of parenchyma at their tips. Roots in this genus depart either directly from the xylem cylinder or from the lateral margins or back (abaxial) side of the leaf traces.

Lunea

The method of leaf trace development in *Lunea* is closer to members of the Osmundaceae than to *Guairea*. The stem of Lunea is sometimes branched and, like *Guairea*, surrounded by leaf traces and adventitious roots (Fig. 3). Its pith and cortex are parenchymatous, each containing clusters of sclerenchyma. Roots arise from the back (abaxial) side of the leaf traces, a feature that is common among members of the Guaireaceae but rare among other ferns. The roots then divide further, producing many additional roots as seen in the cortex.

Donwelliacaulis
The xylem cylinder of the stem of *Donwelliacaulis* is unusual in its composition. It is a tubular mass of several hundred netlike xylem bundles, rather than an intact xylem cylinder like *Lunea* or *Guairea*. These xylem bundles are irregularly shaped in cross section. Leaf traces, also irregularly shaped, originate from adjacent xylem bundles near the outer edge of the xylem cylinder. Initially, the traces form oblong masses (Fig. 5), but eventually become thin and omega-shaped near the outer margins of the cortex.

Itopsidema
The vascular cylinder of Itopsidema is without leaf gaps. Its parenchymatous cortex has numerous leaf traces and adventitious roots, and its pith contains rare, isolated, tracheids among its parenchyma cells. The leaf traces are oblong to C-shaped near the xylem cylinder, becoming horseshoe-shaped near the cortex margin. The roots originate on the outer edge (abaxial) of the leaf traces. The petioles are oblong in cross section without stipular wings, but, like the stem, having spines along their surfaces. Mucliage ducts also occur in the cortex of the petioles. The mushroom-shaped petiolar vascular strand is surrounded by a sheath of sclerotic cells. Insect borings containing very small fecal pellets (coprolites) also occur in the petioles and adventitious roots of Itopsidema. They appear to have been produced by orbatid mites.
The Guaireaceae is a small, but intriguing family. The anatomical structure of its members is rather unique and certainly different than the internal structures in the majority of the other ferns, extant and/or extinct. Further worldwide collecting will eventually uncover other genera that will be included in this family.
It will be interesting to see how large this family becomes, although is not likely to attain the numbers of the closely related family, the Osmundaceae.

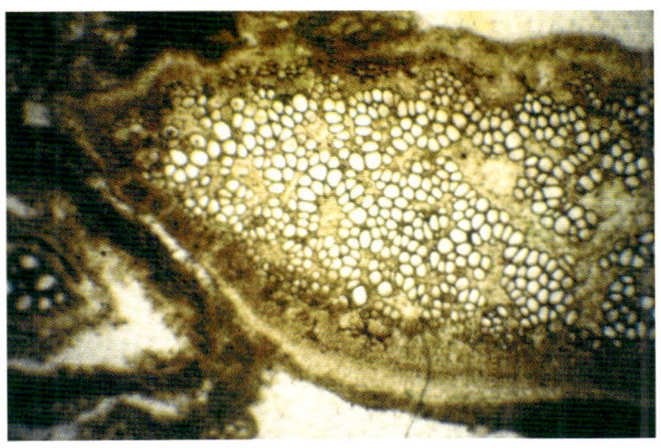

4 A petiole and root (lower left corner) of *Donweliiacaulis chlouberii*

5 *Donwelliacaulis chlouberii*
Note the xylem strands in the petioles

WILLIAM D. TIDWELL

Über den sonderbaren Farn Tempskya aus der Kreidezeit

Tempskya - a Unique Fern From the Cretaceous

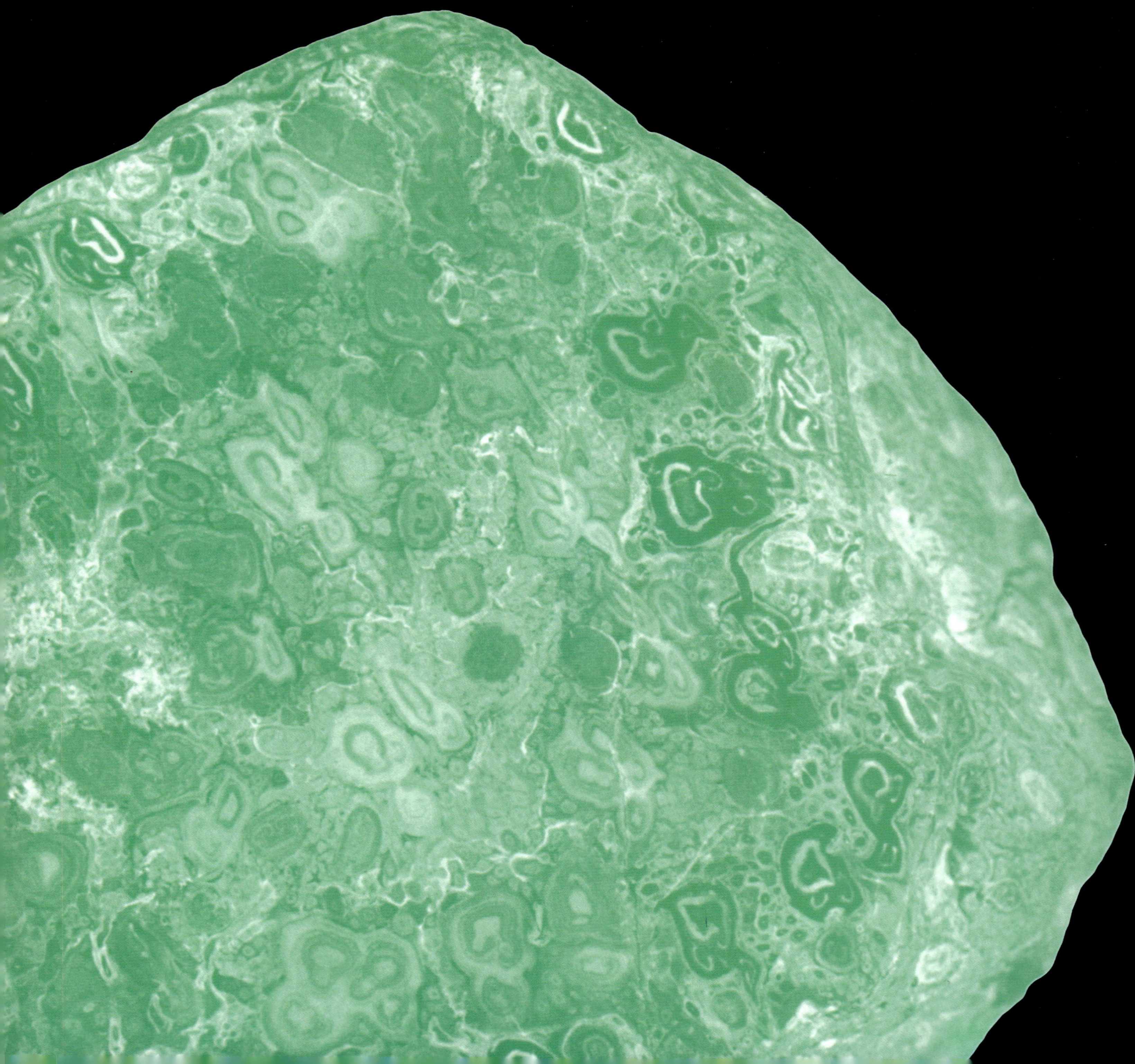

The Cretaceous fern genus, *Tempskya*, the only genus in this family, is of exceptional interest. It occurs as silicified trunk-like structures (false trunks) that may be 0.3 m (12 inches) or more in diameter and upwards of 3.7 m (12 feet) high. The axes are most often weathered and generally collected as small to large pieces. When intact, the trunk-like structures or false trunks are often club-shaped, conical, straight or slightly curved (Fig. 1). In transverse section, they are round or irregularly oval to elliptical. This is due mostly to weathering. The false trunks may resemble palm axes, with which they are often confused (Fig. 6). Internally, the false trunks consist of many branched stems held together by a mat of adventitious roots (Figs. 2-4, 11). The individual stems or branches are small, amphiphloic siphonosteles (solenosteles), that is stems where the xylem is surrounded by phloem, both of which encloses the pith (Fig. 5). Outside of the outer phloem is the pericycle, endodermis, and a three-, rarely two-, layered cortex. The outer cortex is parenchymatous, the middle cortex sclerenchymatous, and the inner either parenchymatous, sclerenchymatous, or a combination of the two, depending on the species.

The leaf traces arise on one side of the stem in two rows or ranks. In cross section, their place of origin is marked first by a thinning of the xylem, and the leaf trace then begins bulging outward. As the bulge expands outwardly, the C-shaped leaf trace separates from the stem first on one side and then the other (Fig. 7). This leaves a leaf gap in the stem,

1 False trunk of *Tempskya minor* in upright growth position (arrow) embedded within sandstone beds east of Castle Dale, Utah

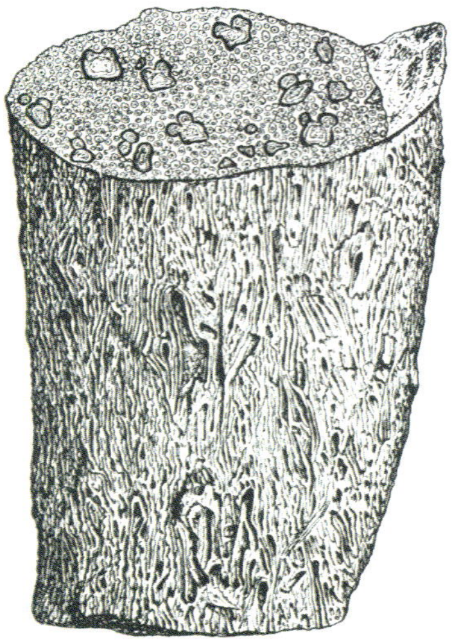

2 *Tempskya knowltonii* illustrating stems within a mass of adventitious roots; note the corrugated nature of the false trunk original size, modified from Brown 1936

which, in turn, gives the stem a horseshoe-shaped appearance before the gap closes (Fig. 5).

The roots arise directly from the stem (adventitious) usually opposite the leaf trace formation (Fig. 5). The roots then extend downwardly toward the base of the false trunks, providing support for it. The xylem elements of the roots are generally arranged approximately in the shape of a "cross" (Fig. 8). This is an identifying character of *Tempskya* roots. Sheathing parenchyma enclose the xylem, as does the phloem, pericycle, and endodermis, if preserved. The cortex of the root is composed of a thick to thin parenchymatous layer and generally a thick fibrous (sclerenchymatous) inner zone, so that the roots appear to have been tough and wiry. Root hairs extend from the epidermal cells, but are often not preserved.
The orientation of the *Tempskya* stems as to their leaf trace departure furnishes a clue as to their possible growth habit. In most species, if not all, the leaf traces depart on the side nearest the edge of the false trunk and generally away from its center (Fig. 3). This is referred to as radial symmetry and indicates an upright growth habit. The second group was postulated to have traces departing in one direction and thus suggesting a reclining or dorsiventral habit. Theoretically, these axes were vine or liana-like. However, there is a strong probability these are merely fragments of the typically radial type and essentially all tempskyas grew upright. Numerous small, kidney-shaped fecal pellets from some unknown animals occur in tunnels excavated into the cortical regions of some stems (Fig. 9), petioles, and roots, and in the masses of epidermal hairs associated with these structures in many specimens of *Tempskya*. Generally the pellets are loosely packed often filling the galleries or tunnels and appear to have been deposited as the animals were feeding upon the thin-walled parenchyma cells and epidermal hairs.

A variety of fungal and fern spore types occur in macerated material from the false trunks of *T. jonesii* from

Utah. The fern spores differ from those described by Boodle in a specimen of *T. schimperii* collected near Brightstone on the Isle of Wight, Great Britian. The spores from the Utah and Idaho (*T. wesselii*) specimens, for instance, are smooth walled rather than being sculptured as are those from *T. schimperii*. Because of the manner in which the false trunks of *Tempskya* grew, it is very likely spores from other types of plants were blown in and preserved in the trunks of this genus, hence the discrepancy. Sporangia also found in the Utah specimens have vertical annuli (Fig. 10) which are different than the apical annuli of the Schizaeaceae and the oblique annuli of the Gleicheniaceae, two families to which some authors have suggested *Tempskya* may belong.

Twelve species of *Tempskya* have been reported from North America; eight, possibly ten, from Europe, most of which are imperfectly known; one from Japan, and the first from Argentina (Figs. 3, 11). As pointed out by Arnold in 1958, *Tempskya* false trunks are rarely found in place, which has introduced some uncertainty as to the stratigraphic range of

3 *Tempskya* sp.
from Neuquen, Argentina (xs)
the first *Tempskya* from the Southern Hemisphere
original size

3a Detail of stems
Note their branching, 2 x

some species of the genus. The only place *Tempskya* occurs in growth position is near Castle Dale, Utah, U.S.A. in the upper Lower Cretaceous Cedar Mountain Formation (Fig. 1).

The *Tempskya* material of the European species had been removed from their original sites of deposition by erosion. The European tempskyas come from a number of horizons and localities. The first specimens were described from the Lower Cretaceous deposits in Tilgate Forest and the beach near Ecclesbourne, Sussex, England. These were originally named *Endogenites erosa*, implying relationship with the monocotyledons, particularly palms. The generic name *Tempskya* was proposed by Corda in honor of Friedrich Tempsky, based on several specimens from gravels along the Elbe River, in Bohemia, Czech Republic. In 1845 Corda described four species from these specimens, one of which, *T. pulchra*, is considered the type. Tempskyas have also been found in northern France, northeast Spain, near Hanover, Germany, the Peruc-Korycany Formation beds near Lana, west of Prague, Czech Republic, as well as the Isle of Wight, to name a few.
The Russian species, *T. rossica*, was

4 *Tempskya jonesii* (ls)
Specimen illustrating the upward branching of the stems surrounded by numerous roots
original size

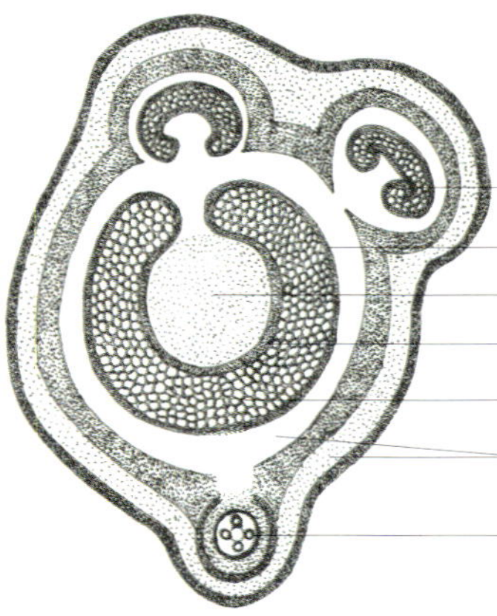

5 Diagrammatic illustration of a *Tempskya* stem showing two-ranked leaf traces and leaf gap

6 Fragment of *Palmoxylon*
0.5 x
modified from Brown 1936

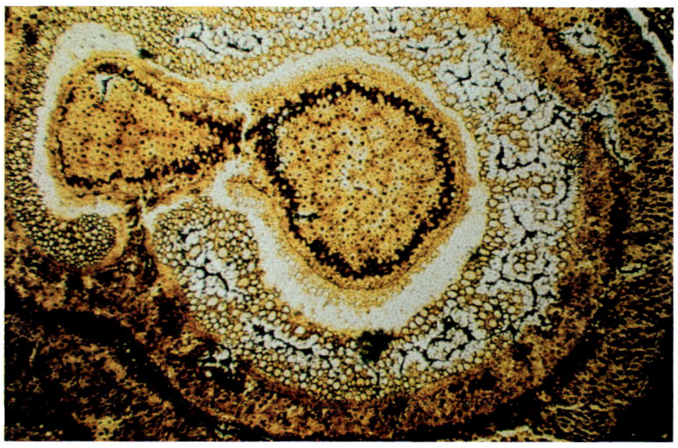

7 Detail of one of the stems in *Tempskya*
Note the departing leaf trace, 25 x

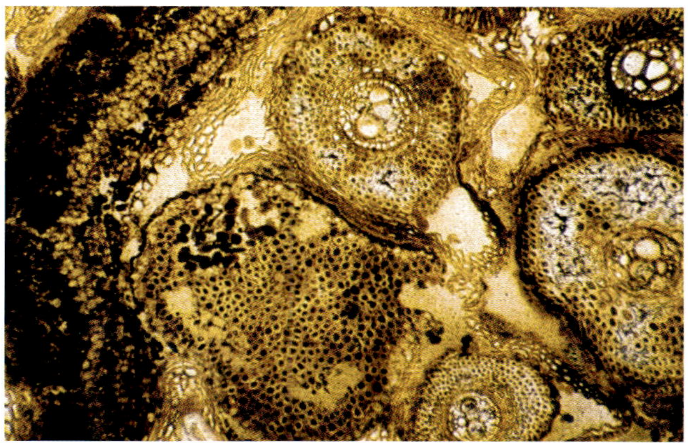

8 Roots of *Tempskya stichkae* invading a stem transverse section, 7.5 x

9 Oblique view showing fecal pellets (arrow) in the inner cortex of *Tempskya* sp. Argentina, 35 x

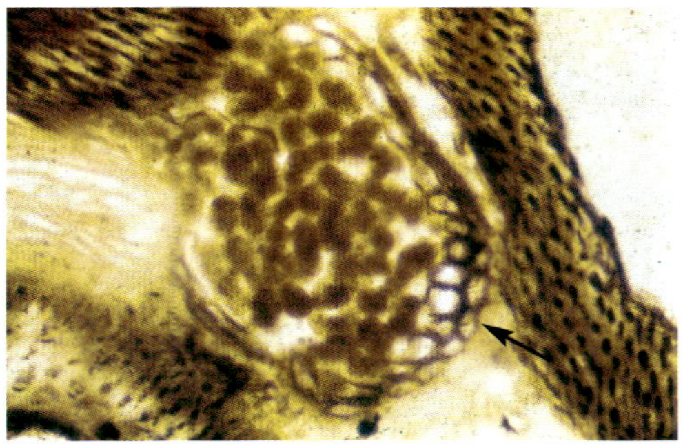

10 Sporangia of *Tempskya jonesii* showing vertical annulus (arrow), 40 x

collected from a Tertiary conglomerate in the basin of the Karaganda River on the west flank of the Mugodjar Mountains, Russia. The specimens were considered to have been reworked from older strata, possibly Upper Cretaceous. Six specimens of *Tempskya* were reported from Quaternary gravels in southern Barvaria, Germany that were thought to have originated from Cretaceous deposits in the Alps (Fig. 12). They were assumed to be possibly related to *T. grandis*, as was the unnamed specimen described from the beach at Sheppey, Kent, England, that was thought to be from the Eocene.
A likely source for the latter specimen is the Lower Cretaceous Lower Greensand, which had also yielded *T. erosa*, one of the first species of *Tempskya* to be recognized.
In the United States, the majority of tempskyas are either reworked in younger gravels or reworked in the formations in which they were originally deposited. With the exception of a few specimens from the early Late Cretaceous (Cenomanian) Dakota Formation, most of the formations in which tempskyas occur in the United States are late Early Cretaceous (Albian) in age.
Tempskya specimens are known to occur in at least 45 localities in the United States. The first American

specimens of this genus were collected from the Patapsco Formation in the valleys of Stony Run and Deep Run north of Severn, Anne Arundel County, Maryland. These were assigned to the species, *T. whitei*. In 1908, a specimen was found in the Musselshell Valley southeast of Harlowton, Montana and described as *T. knowltonii* (Fig. 2). Other specimens of *Tempskya* have been found in the Badlands, northwest of Great Falls, Montana; north of Kemmer, Wyoming; Valley of Beaver Creek, east of Greybull, Wyoming; as waterworn cobbles northwest of Chadron, Nebraska; the Wayan Formation southeast of Grays Lake, Idaho; jasperized material from placer deposits one mile north of

11 *Tempskya* sp. (xs)
Argentina
Dernbach collection, original size

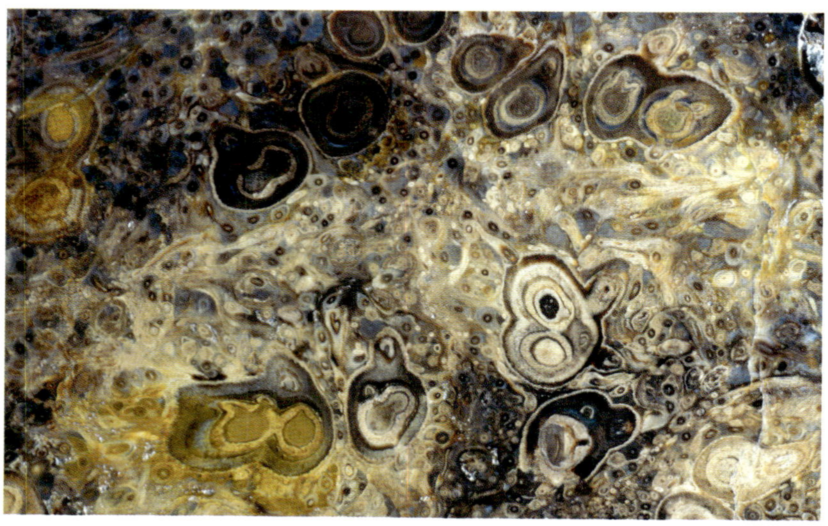

11a Detail of the stems, leaf traces, and roots of Fig. 11, 2.5 x

Fig. 12. *Tempskya* sp. (xs) from Germany
Bavarian State Collection for Palaeontology and Geology, Munich
original size

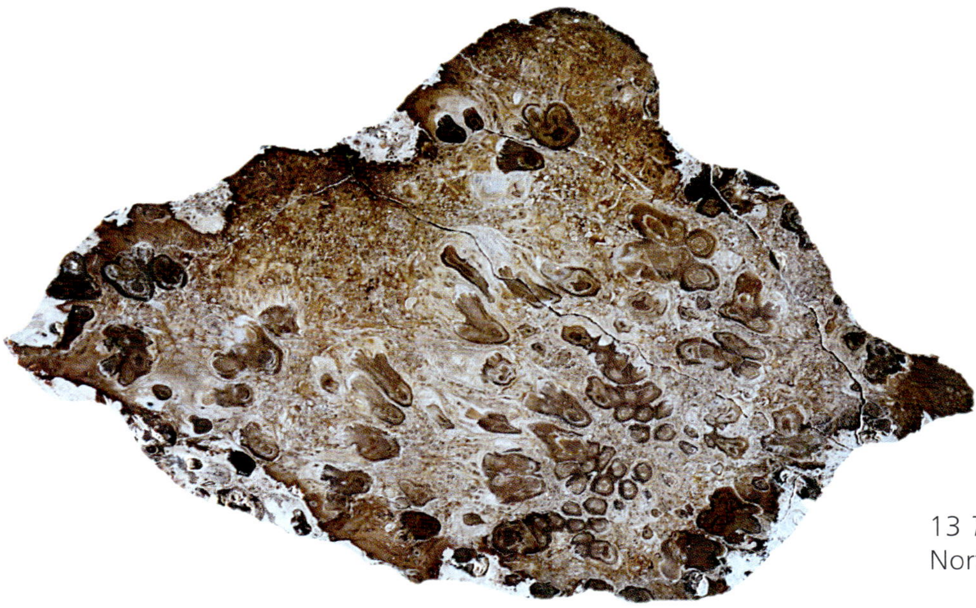

13 *Tempskya* sp.
North Carolina, 0.7 x

Greenhorn, Oregon; in growth position southeast of Castle Dale, Utah near the confluence of Cottonwood Creek and Huntington Creek which then forms the San Rafael River (Figs. 1, 4); east of Ferron, Utah; southeast of Moab Airport, Utah; Blue Mesa site southeast of Moab; in the Willow Tanks Formation southwest of Kaolin, Clark County, Nevada; southwest of Deming, New Mexico; north of Bitlabito Trading Post, New Mexico; and specimens of *Tempskya* come from Quaternary gravel deposits in North Carolina (Fig. 13). The latter is undoubtedly reworked material and occurs with *Palmoxylon* having the same type of preservation. The geologic formation from which they were derived is presently unknown.

ULRICH DERNBACH

Neufunde vom Cerro Cuadrado, Argentinien - Zapfen am Ast

New Finds From Cerro Cuadrado, Argentina - Cones on Their Branches

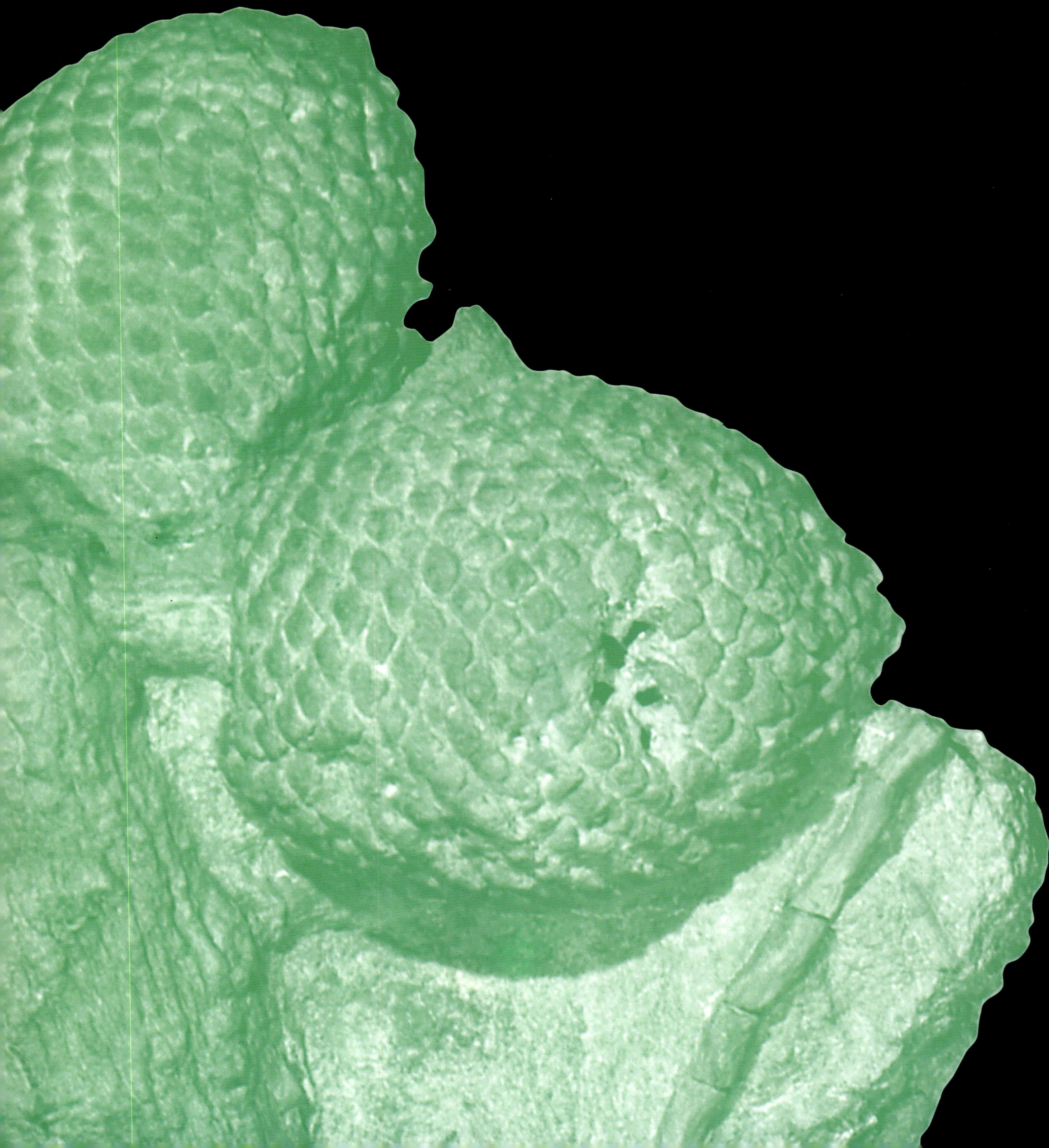

Araucaria cones from Cerro Cuadrado in Santa Cruz province are comprehensively described in my book "Araucaria", where I explain that laborious digging has now become the only way to find additional specimens of these beautiful fossils. The ground at the site is very rocky, which explains why the local gauchos seldom dig deeper than 20 cm.
In 1997, my friend Enrique Alvarez told me about an exciting new discovery. Together with his son, he had dug approximately one meter deeper and found strata in which fossilized araucaria cones were still clinging to their branches. Several such specimens had subsequently been unearthed. In most cases, there are two or three cones on each branch; a greater number of cones per branch is extremely rare. After they're prepared and processed, these specimens will no doubt be very attractive and highly valued as collector's items.

The Invitation

In December 1999, I had the opportunity to respond to Enrique's invitation to join him in a search for buried cones. We left Comodoro Rivadavia very early in the morning and reached Juan Carlos's estancia around midday. The temperature had already risen above 30°C (86°F), but the heat was not unpleasant because the wind, which is such a typical feature of this season, helped to keep us comfortably cool. Our adventure began early the next morning.
Enrique had staked out an area measuring approximately 10 x 4 meters. It was here that we planned to begin digging. The three of us worked very hard during the next three days, and the sweat poured off us in rivers. Ultimately, however, our hard work was well rewarded.
At a depth of slightly more than one meter, embedded in pale tuff, we found several cones still attached to araucarian branches. My particular interest was attracted by a group of cones which were affixed so closely to their branch that they seemed to have been glued there. We then found, and this is the first find of its kind, female cones and some of the very rare male cones of *Pararaucaria* still attached to their branch. We can thus be certain that we are dealing here with monoecious plants, as is the case among many other conifers.

1 *Araucaria mirabilis*
Cones attached to a twig
original size nearly 30 cm
photo: Noll

2 *Araucaria mirabilis*
Cones attached to a twig
original size nearly 35 cm
photo: Noll

3 *Araucaria mirabilis*
Cones attached to a twig
original size nearly 40 cm
photo: Noll

4 *Pararaucaria patagonica* cones
original size, photo: Noll

RUTH A. STOCKEY

Neues über die fossilen „Keimlinge" vom Cerro Cuadrado, Argentinien

A Reinterpretation of the Cerro Cuadrado Fossil "Seedlings", Argentina

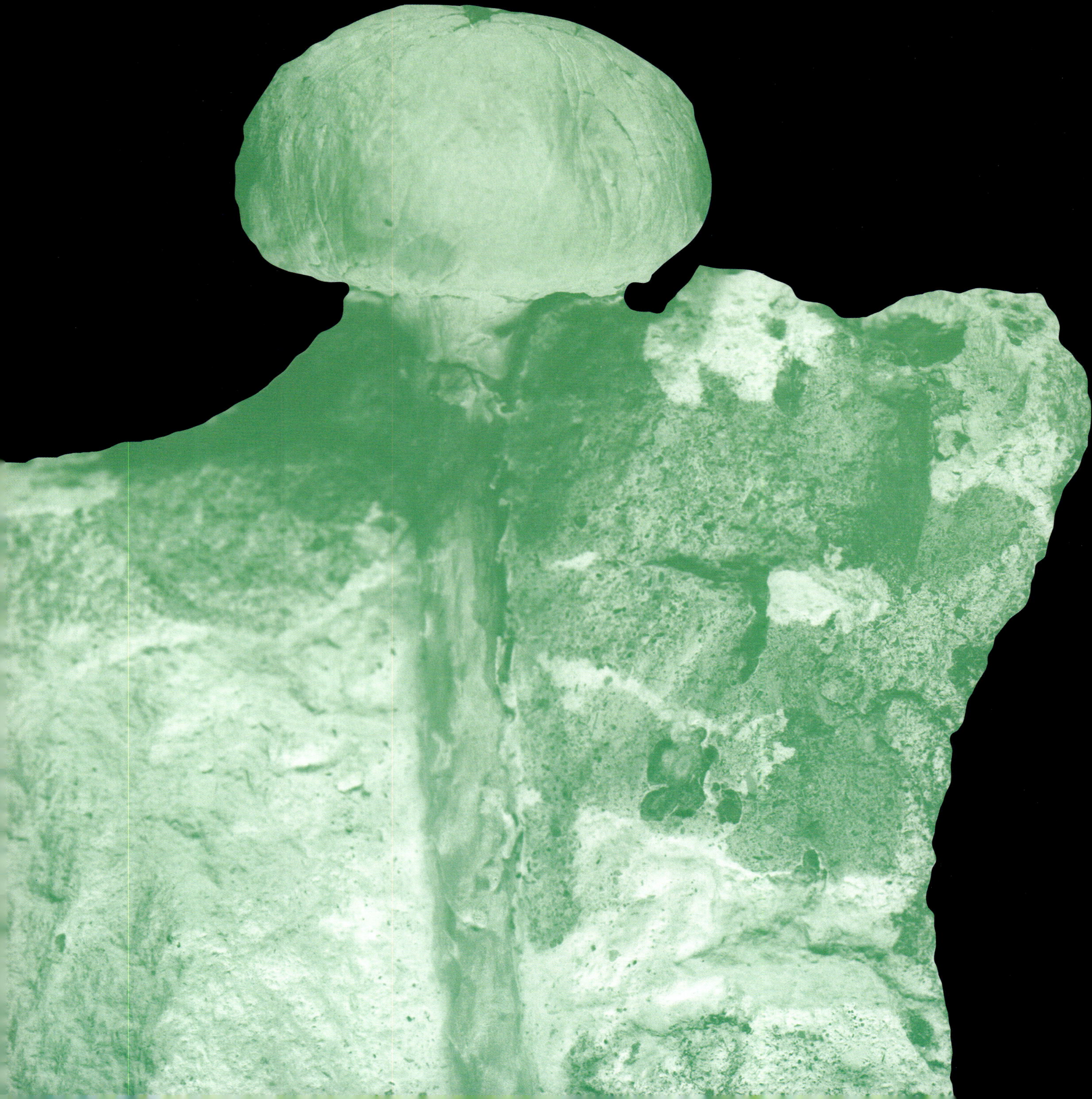

The Cerro Cuadrado Petrified Forest in Patagonia has yielded some of the best preserved fossil conifer remains known. Fossil seed cones of *Araucaria mirabilis* (Spegazzini) Windhausen and *Pararaucaria patagonica* Wieland are some of the best studied fossil remains from these Jurassic conifers. In addition to seed cones, pollen cones have also been described from more poorly preserved specimens. The vegetative remains from the Cerro Cuadrado include, wood, twigs with leaf scars, leaves, and trunks buried in growth position.

Among the most misunderstood remains known from Cerro Cuadrado are what were reported as fossil conifer seedlings. In 1935 Wehrfeld described these as "araucarian fossils" but made no remarks on their morphological nature or structure. In 1950 Gothan originally described what he referred to as "fig-like bulbs" from the collections in Berlin and Stockholm. He suggested that these corm-like or top-shaped structures might actually represent insect or fungal stem galls but described them as decorticated conifer shoots lacking

resin canals. While most authors have opted to call these remains seedlings, it was not at all certain just what the small carrot-shaped, top-shaped, corm-shaped or globular objects might represent.
Part of the problem with the identification of this suite of fossils was that several different types of remains were represented in the Cerro Cuadrado collections around the world. Wieland illustrated what he interpreted to be conifer seedlings in his large publication on the Cerro Cuadrado Petrified Forest. These two elongate structures were reexamined by Stockey and Taylor in 1978 and it was agreed that they did probably represent seedlings as root scars appear at the bottom of each structure and the main axis was composed of conifer wood with uniseriate, circular bordered pits on the radial walls of tracheids. Unfortunately, at the present time it is not certain whether they represent seedlings of *Araucaria mirabilis* or *Pararaucaria patagonica*. The identity of the other "seedlings" were, however, in doubt.

Since fossil *Araucaria mirabilis* was closely compared to the living species *A. bidwillii* Hooker, the Bunya pine from Queensland, Australia, authors naturally compared the corm-like structures found in the Cerro Cuadrado forest to the seedlings of *A. bidwillii. Araucaria bidwillii* produces very large cones and seeds, and seedlings that have hypogeal germination with a large swollen hypocotyl that looks superficially like the more elongated carrot-shaped structures from the Cerrro Cuadrado. In seedlings with hypogeal germination the cotyledons remain inside the seed coats absorbing nutrients from the megagametophyte as the seedling grows. Cotyledons do not emerge from the seed coat and are not photosynthetic. Calder and later Stockey and Taylor examined the more elongated structures and agreed that these could perhaps be the swollen hypocotyls of seedlings. There was no internal anatomical evidence that they were produced by fungi or the result of insect damage. The internal tissue preservation, while not good, looks parenchymatous.
Sections were made of *A. bidwillii* seedlings and comparisons made to the fossils. These authors, however, were not comfortable with the description of all of the fossils from Cerro Cuadrado as probable seedlings. The rounder, more corm-like structures looked very different from the usual *A. bidwillii* seedling.
The turbinate or corm-like structures from the Cerro Cuadrado appear to be different structures from Wieland's two seedling specimens and the more carrot-shaped forms in Stockey and Taylor. These corm-like structures are solid conifer wood composed of short tracheids and narrow vascular rays (Figs. 1-6). They usually show a small parenchymatous pith, lack resin canals, and are decorticated (Fig. 2). The internal tissues are not disrupted in any way and there is no evidence of roots or root traces. One end of each structure is rounded (Fig. 3), whereas the other end may be truncated and extend into an axis (Figs, 1, 4). All parts of these structures are composed of secondary xylem tracheids. Preservation is usually poor, but indications of a growth increment are occasionally visible in these specimens (Fig. 2). Because of the uncertainty of these structures as actual seedlings, several types of investigations have been conducted since 1978 to further aid in our understanding of their structure.

An important discovery of similar "seedling" like structures was made in Cretaceous rocks from Japan. In 1990 Stockey, Nishida and Nishida described these structures in detail. They are similar structures to the turbinate or corm-like bodies present at Cerro Cuadrado. The Japanese material from the Cretaceous of Hokkaido is better preserved and more easily studied using the cellulose acetate peel technique similar to that used for

1 Cerro Cuadrado lignotuber showing truncated end, Dernbach collection, nearly original size

2 Cross section of Cerro Cuadrado lignotuber showing central pith, radiating secondary xylem tracheids and probable growth ring Dernbach collection, nearly 2 x

3 Distal end of lignotuber in Fig. 6 Dernbach collection nearly original size

coal balls (Rothwell, in this volume). The Cerro Cuadrado fossils must be studied using petrographic thin sections. In 1990 Stockey et al. described at least two types of turbinate structures from Japan that differ from those of Cerro Cuadrado. One type has sclerenchyma nests in the pith and lacks resin canals. The second type has a parenchymatous pith and contains resin canals in the secondary xylem. These resin canals appear to be traumatic. In addition, there are two specimens that have resin canals in the primary xylem, and others that have very wide medullary rays. The authors suggest that there may be more than one type of "seedling" with resin canals in the Japanese Cretaceous, perhaps as many as three or four types, all differing from those at Cerro Cuadrado.

This Japanese research thus questioned the identification of the Cerro Cuadrado "corms" as seedlings. Since araucarian conifers normally lack resin canals in the wood, it seemed impossible to equate most of the Japanese fossils with those of araucarian conifers. Stockey et al. suggested that several conifer families may have been producing these structures. They have several features in common:

(1) general shape
(2) size
(3) a composition of almost entirely

4 Lignotuber with two swellings
Dernbach collection, nearly original size

short secondary xylem tracheids and a small pith
(4) endarch primary xylem
(5) sparse, uniseriate circular-bordered pits on radial walls of tracheids
(6) uniseriate rays usually one to six cells high
(7) broad, medullary rays
(8) and one to two ovoid pits in the cross field.
Since the structure of more mature *Araucaria bidwillii* seedlings was not known, further work on this species was initiated. When Stockey and Taylor sectioned seedlings of *A. bidwillii* and *A. angustifolia* (Bertol.) O. Kuntze (species with hypogeal germination and swollen hypocotyls), their results indicated that the internal structure of these was mostly parenchymatous. The hypocotyl is used as a storage organ underground for the starches, etc. that are transferred from the megagametophyte of the seed via the cotyledons causing a swelling of the hypocotyl of the seedling.
Was there any indication of this extensive secondary growth in araucarian seedlings? In 1990, Stockey et al. described one such seedling. This specimen was an abnormal seedling of *A. bidwillii* that had root apex damage. This seedling shows a turbinate or corm-like swelling near the root end, lacks a tap root, but shows several adventitious roots produced to one side. Internally, this structure was composed of secondary xylem. The shoot apex end, however, showed a series of six concentric vascular strands, unlike anything seen in any of the fossils from Japan or Argentina.
Did root or shoot apex damage produce the structures seen as fossils? The Cerro Cuadrado fossils were the result of burial in volcanic ash. Periodic volcanism fossilized conifers in this area at several levels. Stockey et al. suggested that damage to growing seedlings was possible due to possible scorching related to volcanic activity.
Two lines of evidence, however, have suggested that these corm-like structures are not seedlings per se. Firstly, Burrows et al. and Burrows and Stockey investigated living plants with hypogeal germination and the germination and developmental anatomy of *Araucaria bidwillii*, in particular. In Bunya pine seedlings at about four years of age the corm-like tuber, indeed can show large amounts of secondary xylem similar to that seen by Stockey. But the unusual patterns of xylem formation with several cambia around individual vascular strands seen in *A. bidwillii* are not seen in any of the supposed fossil seedlings. Further doubt as to

the seedling status of these fossils was suggested by Burrows et al. noting Clarkson and Clifford's study of cryptogeal (=hypogeal) germination. Two main characters are associated with this type of germination: fusion of the cotyledons and large-seededness. Bunya pine has these characters, however, *Araucaria mirabilis* does not. Seeds of *A. mirabilis* from the Cerro Cuadrado are small compared to those of living cryptogeal species and their cotyledons do not show fusion.

A seedling the size of the fossils seen at Cerro Cuadrado could not have been produced by a seed the size of *Araucaria mirabilis*. By 1992, we knew that these structures were probably not seedlings. But what were they?

The second line of evidence comes from the Jurassic Morrison Formation of Utah, Wyoming and Colorado. Structures similar to the elongated "seedlings" were found attached to coniferous shoot systems and described as *Behuninia*. *Behuninia* short shoots are usually weathered and preserved as casts. When they are anatomically preserved they have a pith, primary and secondary xylem and occasionally cortex and leaf traces. While these structures look similar to the elongated "seedlings" from Cerro Cuadrado morphologically, the internal anatomy of these structures differs from those in the Morrison Formation. Tidwell and Medlyn compare *Behuninia provoensis* (Chandler) Tidwell and Medlyn to two of the decorticated shoot-like structures described from the Cerro Cuadrado. In my opinion, these two specimens, like *Behuninia*, do appear to be shoots.

The view that many if not all of the structures seen in the Cerro Cuadrado are decorticated shoots goes back to Gothan. While living Araucariaceae do not show short shoots, it is possible that *Pararaucaria*, the other common conifer at Cerro Cuadrado, did. *Pararaucaria patagonica* has no affinities to the Araucariaceae, but most likely has affinities to Pinaceae, Taxodiaceae or Cheirolepidiaceae. Permineralized cones similar to *Pararaucaria* have been found in the Upper Jurassic Morrison Formation sediments (Tidwell, personal communication) but have not so far been studied.

Recent finds from the Cerro Cuadrado Petrified Forest in Patagonia (Figs. 1-6) indicate that these structures are attached to a shoot system (Figs. 5-6). The specimen illustrated in Fig. 5 shows a decorticated branch with helically arranged leaf traces and what appears to be a terminal, corm-like, woody structure. A second specimen shows a larger branched twig with a terminal large corm-like structure (Fig. 6). Small branches are attached to this larger twig. Anatomy of the large specimen shows typical conifer wood lacking resin canals with a parenchymatous pith and narrow vascular rays. A third specimen is very unusual in the Cerro Cuadrado collections in that it shows two closely spaced swollen areas (Fig. 4). Clearly these areas of anomalous secondary growth are not seedlings, and the corm-like structures are not parts of a long shoot/short shoot system as in *Behuninia*. They are, however, attached to shoot systems. Tidwell and Medlyn further describe specimens from the Jurassic Morrison Formation referred to as *Steinerocaulis*. These specimens appear to be similar to the corm-like structures from the Cerro Cuadrado. Most are found isolated, but some appear to be attached to a twig. In addition, they report two structures similar to the corm-like structures from Cerro Cuadrado. These appear to me to be similar to the structures described as *Steinerocaulis* and similar to the structures illustrated here.

Recently conifers and *Ginkgo* have been shown to produce what are referred to as lignotubers. Lignotubers have often been called "basal chichi" or "zhong ru" in *Ginkgo* or "burls" in conifers. They were first thought to be galls; however, we now know that they occur as part of the normal

5 Lignotuber at end of small decorticated twig with leaf trace scars
Dernbach collection, nearly original size

ontogeny or development from buds located in the axils of cotyledons. They can be stimulated to grow by some traumatic event and serve to regenerate the plant.
Ginkgo can sometimes produce induced aerial lignotubers along the trunk and branches associated with a bud. These structures can produce roots and shoots on contact with the soil. Similar structures have also been reported in *Sequoia* following traumatic injury such as fire. They form from axillary buds on young branches. This scenario seems the most likely to describe the Cerro Cuadrado corm-like structures. Burls have also been reported beneath the bark of trees of *Araucaria bidwillii* (personal observation) and trunk burls are known to occur in *Sequoia*. While most of these large burls do not show the form of rounded corms with truncated ends, as in most Cerro Cuadrado specimens, it is possible that the upper branches of older araucarian trees, like those of *Sequoia* can also produce lignotubers in their leaf axils. It is interesting to note that one specimen illustrated by Stockey and Taylor showed a root-like end and a small branch on one side. This specimen I would now interpret as a lignotuber that had touched the ground and begun regeneration.
The area of the Cerro Cuadrado Petrified Forest has repeated volcanic sequences. It is probable that the conifers that grew here were regenerating by means of aerial lignotubers. While these structures acted like seedlings in the regeneration of new plants, they probably arose in the axil of a leaf and reached sizes of at least 6 cm based on all of the collections that I have observed. The large numbers of these lignotubers indicate that their occurrence was common in the conifers that produced them. Furthermore, the presence of several types of these structures in the Cretaceous, Japanese collections and the Jurassic Morrison Formation of Utah and Wyoming indicates that several different conifers may have had the ability to produce lignotubers in the past. These structures are probably partially

responsible for the continued survival of *Ginkgo biloba* and perhaps *Sequoia sempervirens.*
After more than 75 years of speculation we have probably arrived at an understanding of most of these anomalous fossils. It seems clear that at least three types of structures are present at Cerro Cuadrado that have been lumped into the category of "seedlings". The Cerro Cuadrado collections contain true woody seedlings, decorticated shoots, and turbinate structures now interpreted as aerial lignotubers. The latter may have arisen in the axils of leaves in response to injury, probably the result of volcanism accompanied by fire. While most specimens collected over the years are well preserved externally, the large collections made in the 1920's (Field Museum of Natural History, Chicago; Natural History Museum, London; Naturhistoriska Riksmuseet, Stockholm) show evidence of fire scorching of cones and twigs. These are the specimens that are seldom illustrated but that provide important ecological information. As Del Tredici in 1998 points out, lignotuber producing species are most commonly found in ecosystems that are characterized by hot, dry summers and periodic fires.

The rapid burial of plant parts in volcanic ash at the Cerro Cuadrado in Argentina has enabled the study of many plant parts, some found in attachment. The large scale removal of spectacular fossil cones, however, has left us without an appreciation for the placement of these specimens within the sediments. Nevertheless the large numbers of fossils recovered has added to a more complete picture of the Jurassic vegetation. With further investigations of more complete Cerro Cuadrado specimens such as those illustrated in this paper we may finally be able to reconstruct the conifers found here as whole plants and come to an understanding of their development, reproductive biology and ecology.

6 Large lignotuber at end of branching decorticated twig
Dernbach collection, nearly original size

ULRICH DERNBACH
WALTER JUNG

Die Riesenzapfen aus Chubut, Argentinien

The Giant Cones From Chubut, Argentina

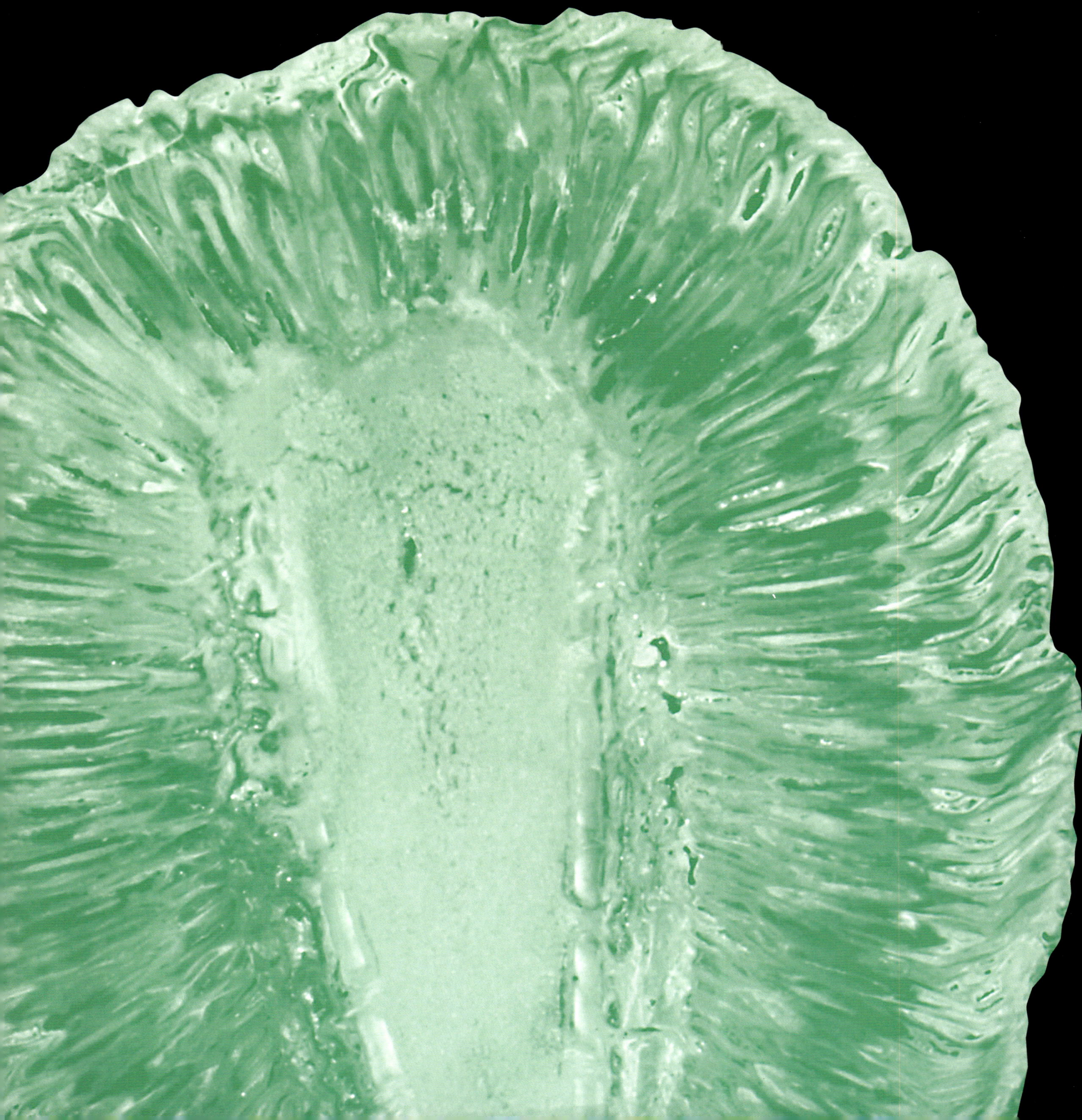

Fossil conifer remains are found in many geological formations and at many locations throughout the world in strata dating from the Late Carboniferous onwards. In addition to nearly uncountable quantities of ancient pollen, the majority of these fossils consist of more or less well-preserved wood and twigs. Specialists devote particular attention, however, to discoveries of seed cones because these fossils can provide extremely important clues for the identification of the specimens. If these cones are preserved not just as encoaled remains or as mere imprints, but consist of well-preserved, petrifactions, then palaeobotanists' joy is truly unbounded.
In such cases, it is often possible to go beyond mere descriptions of external features and to gather a great deal of information about its internal structure. The discoveries of araucaria cones in Patagonia are therefore a case of marvelous good fortune.

In the years following their first discovery by the geologist Windhausen in 1919, such cones gradually found their way into countless private and public collections and museums.
It was rather exciting when, at the Munich Mineral Fair in 2000, a cone in rhyolitic matrix was offered for sale which, even at first glance, attracted immediate attention simply because of its uncommonly large size (length: 12 cm, diameter: 8 cm). Furthermore, this cone had not been found in the province of Santa Cruz (where the old, familiar cones of *Araucaria mirabilis* are found), but further north, in the neighbouring province of Chubut. Some time earlier, U. Dernbach had heard about the giant cones of Chubut from Argentine collector Enrique Alvarez. Now that Enrique was willing to make a few pieces from his collection available for palaeobotanical processing, there were no further obstacles to a scientific investigation of the specimens.

1 E. Alvarez and U. Dernbach examining a giant cone

The "New Ones" (U. Dernbach)
My Argentine friend E. Alvarez first informed me in 1999 about a new discovery of remarkably large and heavy cones. Some of these giants, he told me, were more than 20 cm in length and as much as 8 cm in diameter. A few of the colossal petrifactions, he claimed, weighed more than 2 kg. Naturally, I was very curious about them. My first opportunity to hold one of these specimens in my hands came in 2000. I was truly impressed: Enrique hadn't exaggerated. These cones were simply unbelievable. Because the dimensions of these newly discovered specimens differed so drastically from the sizes of the familiar araucarian cones from Santa Cruz, it seemed quite likely that these specimens must be remains of an entirely different cone. Even before I embarked on my expedition to Patagonia, palaeobotanist W. Jung of Munich had already assured me that he would be glad to study and identify them. E. Alvarez generously gave me several specimens for this purpose. His willingness to part with these items deserves all the more praise because Enrique himself is an impassioned collector.
Of course, I was extremely curious to find out what lay hidden within the cones. The first cut specimens were extremely attractive. Though significantly darker than their famous relatives from Santa Cruz, they compensated for their dusky hue by having lovely blue chalcedony filling inside their seed cavities, as well as some small, isolated, reddish druses. The task of grinding and polishing the cut faces proved to be rather time-consuming because the petrifactions contained a great deal of iron. Enrique had said that these cones had been found in Chubut, in deposits dating from the Early Triassic. But "Chubut" as the alleged discovery site was too imprecise to satisfy me, and I was also quite skeptical about the assigned age, especially after consulting a geological map of Patagonia confirmed that extensive parts of Chubut should be assigned to the Cretaceous.

At the Discovery Site
It was, therefore, indispensable to know the discovery site firsthand.
A long-awaited opportunity to do so finally arrived in May 2001. Having taken off from Buenos Aires a few hours before, my plane landed in the early morning hours in Neuquen, where Enrique was awaiting my arrival. From here we drove past Bariloche, the so-called "Switzerland of

2/3 Medial longitudinal sections of two seed cones
of *Araucaria alvarezii* showing cone axis and attached scales
the interior of the seeds is completely preserved by chalcedony
Dernbach collection, length: 14 cm and 16.5 cm, photo: Noll

4/5 Landscape around Neuquen

6 The largest seed cone of *A. alvarezii* found thus far, Dernbach collection length 19 cm, photo: Noll

Argentina", on our way to Esquel, an idyllically situated vacation spot in Parc National Los Alerces. During the night, the temperature fell to several degrees below freezing. After all, it was now early May and the Southern Hemisphere's winter was fast approaching. We left Esquel very early the next morning, drove about 6 miles towards El Bolson, then turned off route no. 40 onto road no. 12 and headed towards Paso del Sapo. The gravel surface of the road slowed our progress somewhat, but after a few hours, we duly arrived at the shore of Rio Chubut with its lovely, deep-blue waters. When we reached the mighty cliff of Piedra Parada around noon, we knew that only about one more hour separated us from our destination: the site where the giant cones had been found. When we arrived there, I saw that the discovery locality was limited to an area of less than one square kilometer. The relatively small size of this site, Enrique explained, was one of the reasons why only a few cones had thus far been found. We spent several hours at the site. I took careful readings with the GPS and afterwards had time to go "fossil hunting". My joy was great when I discovered a nearly perfect cone, complete with its stalk. This cone measured a stately 19 cm in length (Fig. 6).

Afterwards we reluctantly left the site, and by around 10 p.m. we had again returned to our lodgings in Esquel. Shortly thereafter, when I compared my GPS reading with the coordinates on my geological map, I wasn't particularly surprised to find that the cone discovery site is indeed located in Cretaceous strata. When I returned to Germany, I found the scientific report about the giant cones awaiting me.

What Science Has to Say (W. Jung)
The scientific evaluation is based on examination of several dozen specimens, as well as several fragments

7 Rio Chubut

8 At the fossil site

9 Enrique at work

11 Typical form of an *A. alvarezii* seed cone
Bavarian State Collection, Munich
length 12 cm, photo: Bergmeier

10 Folded exterior portion of a seed cone
of *A. alvarezii*, Dernbach collection
width 2.5 cm, photo: Noll

which often show details more readily than can be seen on intact, uncut examples. With very few exceptions to which the surrounding matrix still adheres, the cones from Chubut were isolated pieces. The specimens I studied were usually in the form of medially sawn and polished halves, as well as a number of specimens which remained uncut and in the same condition in which they had been found. Two cones were tangentially cut and polished; one cone was cut transversely and polished.

In general, the cones' outward appearance suggests that their state of preservation ranges from satisfactory to very good. Unlike the cones of *Araucaria mirabilis* from Santa Cruz, however, the inner fine structure of the Chubut specimens is only moderately well preserved. As a result, a number of questions remains unanswered. This situation could have something to do with the degree of maturity that the cones had attained at the time of their petrifaction. In none of the Chubut cones was it possible to observe an embryo or even a clearly formed "primary endosperm." The parts of the tissue near the cones' peripheries (cone scales or bracts) are often severely flattened and folded, which means that they were obviously still soft and only slightly lignified. Some specimens have been subjected to wind erosion, consequently they have lost part of their exterior tissue. These fossils, however, reveal the best views of the cones' interiors.

The View From the Outside

As previously mentioned, one of the first differences one notices when comparing these cones with those of *Araucaria mirabilis* is the larger size of the specimens from Chubut. The literature mentions mirabilis cones ranging from 2.5 to 9 cm in length and with a maximal diameter of 9 cm, whereas the cones from Chubut (which will be named *Araucaria alvarezii*, in honour of their discoverer) reach a length of 20 cm with diameters ranging from 7 to 11 cm. The average length of an *A. alvarezii* cone is not quite 15 cm.

A second and no less obvious difference between these two cones involves their disparate forms. The majority of the mirabilis cones are nearly spherical, short elliptical or slightly oval; the *alvarezii* cones, on the other hand, are inverted ovoid with the larger width in the upper half of the cone; in rare cases they may be broadly elliptical to spherical.

Another very distinct difference involves the shape of the cone scales, or rather the "double scale complexes". An upwardly bent, sharp, triangular point can be observed on the scales of well-preserved specimens of *A. mirabilis*, although this feature is lacking from the majority of fossils. Where absent, one sees instead the "ligula" as the free end of an outgrowth of the cone scale. In specimens of *Araucaria alvarezii*, the bract likewise culminates in a sharp (albeit narrower) point which, however, is turned obviously downward towards the base of the cone. In this way, it offers the observer an unimpeded view of the "ligula", which is rhombic to triangular in cross-section.

The aforementioned characteristics are not immediately apparent on cones whose exterior surfaces have undergone a greater amount of weathering.

Corresponding to their larger dimensions, the axes of the Chubut cones bear at least twice as many scales (an estimated number would range between 800 and 950) as are borne on the axes of *A. mirabilis*, likewise in 21

12 Typical shape of an *A. mirabilis* seed cone
Jung collection, length 7 cm
photo: Bergmeier

13 Surface of a seed cone of *A. alvarezii* with well-preserved scales
Bavarian State Collection, Munich
width 4.5 cm, photo: Bergmeier

14 Surface of a seed cone of *A. mirabilis*
with well-preserved scales
Bavarian State Collection, Munich
width 5.5 cm, photo: Bergmeier

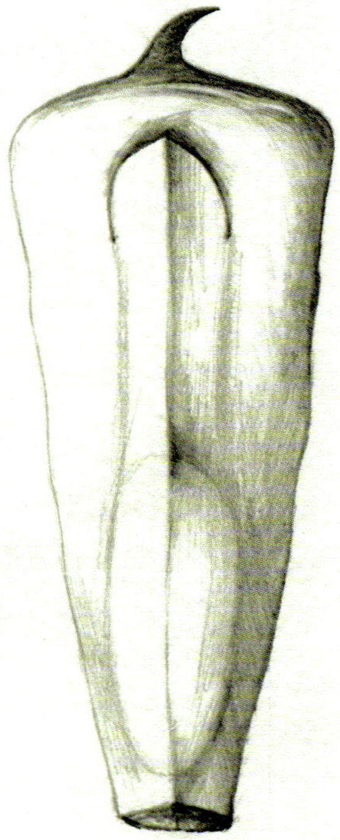

16 Reconstruction of a double-scale complex of *A. alvarezii*, drawn in a scale of 3 to 1 by J. Jung

15 Surface of a slightly rolled cone of *A. alvarezii*
The bracts and their underlying ligules are clearly distinguishable from one another.
Bavarian State Collection, Munich width 6.5 cm, photo: Bergmeier

parastichies. Additional specimens will be required before the position of the cones on the twigs can be determined.

The Interior of the Cones

Insufficient preservation and irregular silicification of some specimens makes it impossible to discern all of the details of the cones' construction. It is clear, however, that the structure is composed of "double scale complexes." Each of these complexes consists of a large covering scale and a much smaller seed scale which is fused to its covering scale. The distal part of the seed scale, which is not fused with the covering scale, is known as the "ligula." The cone scales are one-seeded. The thick-walled, mature seed is enclosed within scale tissue. The aforementioned relationships correspond to those that have already been observed in *Araucaria mirabilis*, so that it is possible to assign the Chubut specimens to this same genus.

Largeness is the salient characteristic of *A. alvarezii*, both in the dimensions of the entire cones and in the dimensions of the individual cone scales. The Chubut cones and their scales are longer and broader than those of *A. mirabilis*. In his examination of *A. mirabilis*, Calder reported scale lengths ranging from 13 to 16 mm, with an average width of 10 mm; the corresponding dimensions in *A. alvarezii* scales are 30 to 34 mm in length and 10 to 15 mm in width. Like the *mirabilis* scales, the scales on the Chubut cones are obviously laterally winged. Because of extensive destruction of the tissue, histological details can be reconstructed only in a few cases, thus precluding the possibility of comparison with the relationships in *A. mirabilis* which are so well-known thanks to the work of Calder and Stockey. It was possible to discern up to 12 vascular bundles in the bracts and at least two such bundles in the ligula of *A. alvarezii* scales. The ovules were obviously immature. Neither megaprothallia nor embryos appear to have formed. Only rather vague statements can be made about the possible pathways of the vascular strands. Observations of two cones, however, support the supposition that two separate traces per cone scale occurred at each node in the vascular bundle network. This impression, if verified, would have implications for the relationship between this fossil species and extant araucarias.

Where Are the Relatives ?

Research conducted during the past several decades strongly suggests that *A. mirabilis* is very likely related to *A. bidwillii*, an extant species native to eastern Australia. The current state of scientific research, however, does not provide the necessary data upon which to base similar assertions about the Chubut species. Too many characters and anatomic details remain unknown. First, and as is also the case for *A. mirabilis*, pollen cones have not yet been identified with certainty. Second, the precise pathways and structures of the vascular bundles and resin canals are still unclear. Third, scarcely any data has been gathered about the design of the ovules and no data is presently available about the development of embryos and seedlings in *A. alvarezii*, although the corresponding information is known for *A. mirabilis*. Finally, no leaf-bearing *A. alvarezii* twigs have been discovered. Knowledge derived from imbricated "*Brachyphyllum*" twigs from the Cerro Cuadrado region cannot be transferred to the Chubut cones. If subsequently discovered material verifies the supposition that two separate vascular strands indeed lead into each scale complex, then this fact, together with the presence of fairly massive wings on the cone scales and obvious vascular bundles in the "ligula," would be an impossible-to-overlook clue suggesting that *A. alvarezii*, like *A. mirabilis*, is related to recent *A. bidwillii*. Whether the criteria of differentiation in the recent *Araucaria* section can be applied without restriction to the several species of fossil araucarias is a question which remains unanswered for the time being. In terms of their structural preservation, none of the giant cones from Chubut, including the remarkably large cones of *A. sphaerocarpa*, can compare with fossil araucarian cones found thus far at various localities in England, France, South Africa, Asia, and New Zealand.

17 Wind-eroded surface of an *A. alvarezii* seed cone
The vascular bundle of the bract and its accompanying ligule are visible as pale points.
Bavarian State Collection, Munich
width 3 cm, photo: Bergmeier

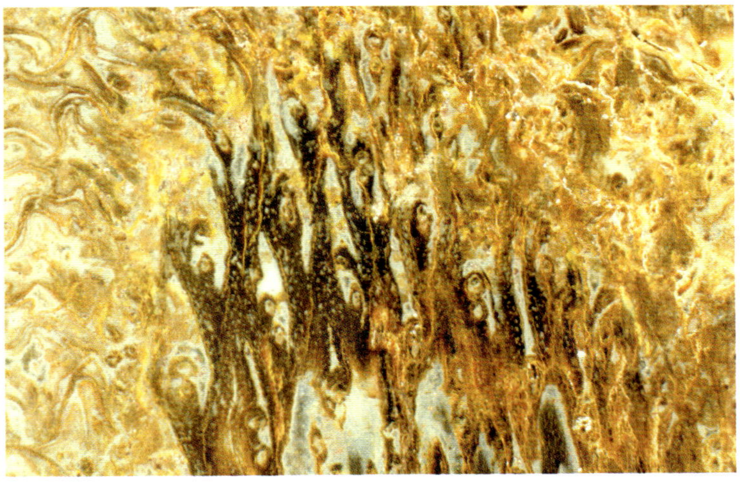

18 Tangential cut through the cone axis of an *A. alvarezii* seed cone; the crosswise egresses of double vascular bundles leading into the cone's scales can be seen within the nodes of the dark net of vascular bundles
Bavarian State Collection, Munich
width 1.6 cm, photo: Bergmeier

ULRICH DERNBACH

Der Kreidewald von Rio Negro, Argentinien

The Cretaceous Forest of Rio Negro, Argentina

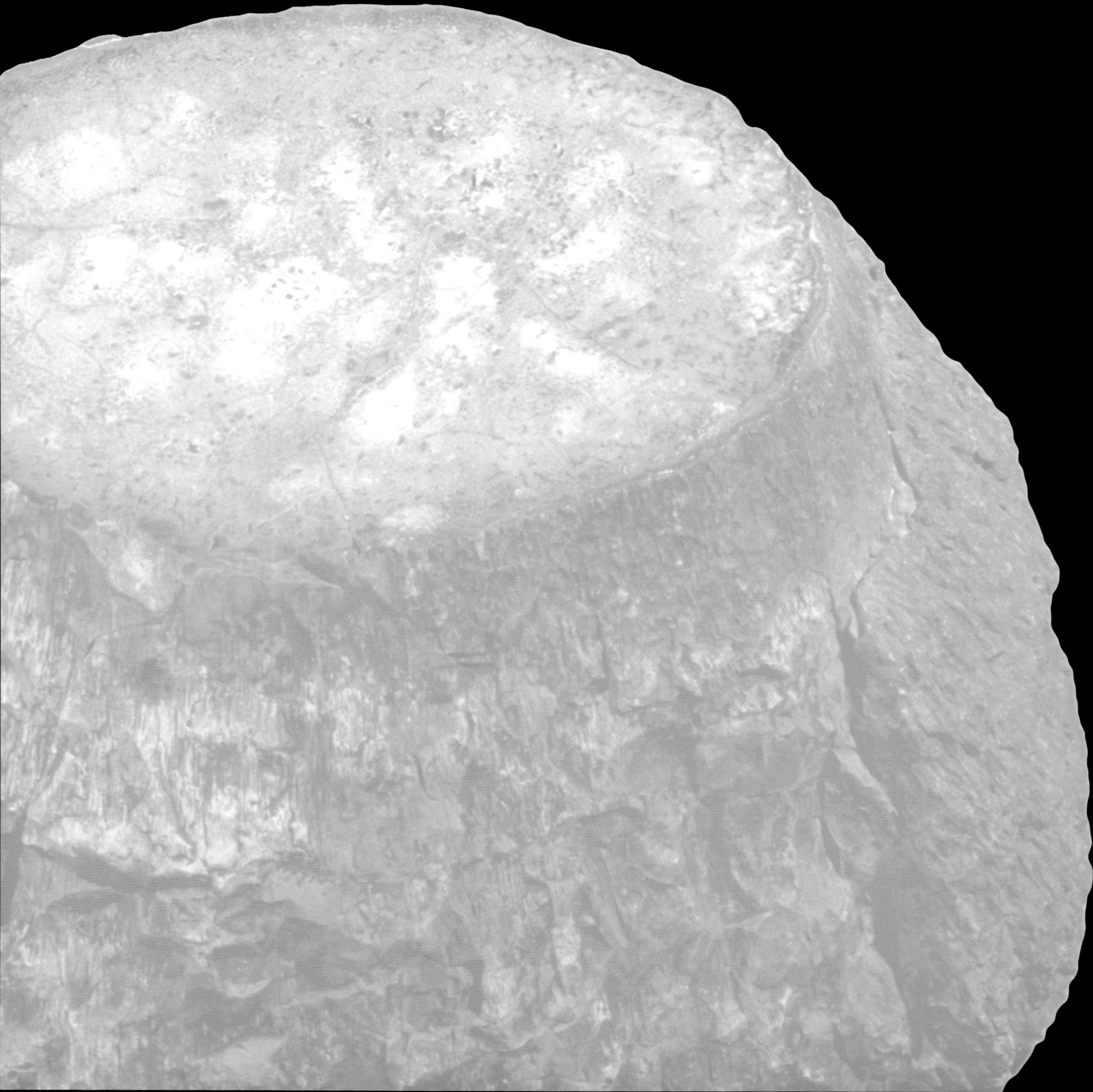

As early as 1969, Bruno Petriella reported the discovery of fossilized cycadophytes in Los Menucos, Rio Negro province of Argentina.
He tentatively assigned these petrifactions to the Lower Tertiary.
In recent years, cycadophytes from Rio Negro have repeatedly been offered for sale at the mineral fair in Tucson, Arizona. The presumed ages assigned to these specimens have varied.
My Argentine friend Enrique Alvarez had already told me in 1999 about the existence of interesting petrified woods in Rio Negro so he and I decided to pay a personal visit to the site. The mercury frequently rises to more than 40°C (104°F) during the summer months, so we scheduled our expedition for November 2000.

The Discovery Site

We left Neuquen and drove eastwards along highway 27. We passed through the little town of Villa Regina, then turned off the main road. After continuing southwestwards for approximately another 100 km, we reached the discovery site. Enrique knew the gaucho and his family from earlier visits, so their reunion was very cordial indeed. While sipping cups of maté, we explained to the gaucho why we had come. Shortly thereafter he freely granted us his permission to search for fossils. Our adventure had begun! The area where we found fossilized woods during the next three days stretches across several square kilometers. It became clear to us that the majority of specimens had been em-

1 Lavishly colored cycadophyte
Collection of the Dr. Haas family
width 22 cm, photo: Noll

2 The Cretaceous forest of Rio Negro

3 Land turtle
Chelonoidis chilensis donosobarrosi

5 Detail of Fig. 4
8.4 x 5.6 mm
photo: Noll

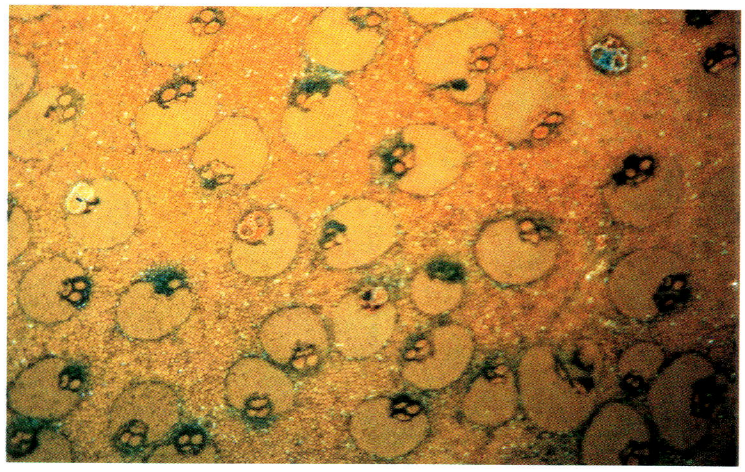

4 Beautifully colored palm
17 x 14 cm, photo: Noll

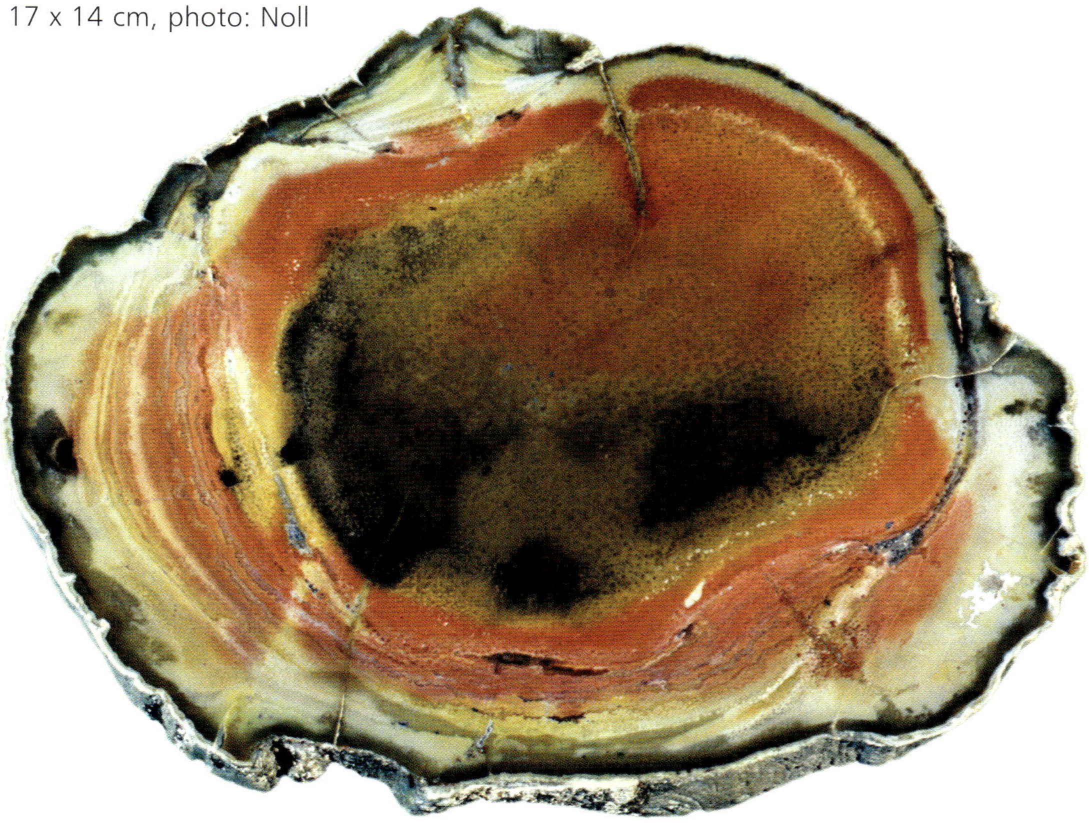

6 Lavishly colored araucaria
14 x 11 cm, photo: Noll

7 *Araucaria* with brightly colored agates
24 x 18 cm, photo: Noll

8/9 *Araucaria* in transverse section
with brightly colored agates
18 x 14 cm, photo: Noll

10 Dinosaur bone filled with agate
Mayer collection
7.5 x 3.5 cm, photo: Noll

bedded in sandy sediment and that their longitudinal axes were almost always in a northwest-southeast orientation. This suggested that they had been deposited here by floodwaters. Unfortunately, the literature includes no mention of this discovery site in Rio Negro, so it was impossible to make any definite statements about the stratigraphy. When I compared the readings that I had taken with my GPS with the coordinates on a geological map of Patagonia, I was somewhat surprised to discover that we were searching for petrifactions in Cretaceous deposits.

Evaluation of the Discoveries

We spent a total of three days at the Cretaceous forest of Rio Negro. Our "harvest" was quite abundant. Red and yellow colored wood from ancient palms and conifers with excellent agate formations were among the most frequent finds. Rarer specimens included strongly quartzed, red-blue-yellow colored pieces of cycadophytes. Unfortunately, we were not able to find a complete cycadophyte trunk (Tidwell, in this volume).
In addition to fossil woods, there were also a large number of fossilized dinosaur bones which, like the vegetative remnants, almost always included perfect agates.

Summary

A mere three-day sojourn at this discovery site is far too brief. Additional visits in coming years will be indispensable. I'm sure that this Cretaceous forest still contains many undiscovered mysteries.

11 Stem surface of *Cycadeoidea*
width 10 cm, photo: Noll

WILLIAM D. TIDWELL

Die Cycadophyten - ein Gruß aus dem Zeitalter der Dinosaurier

The Cycadophytes - Greetings From the Dinosaur Age

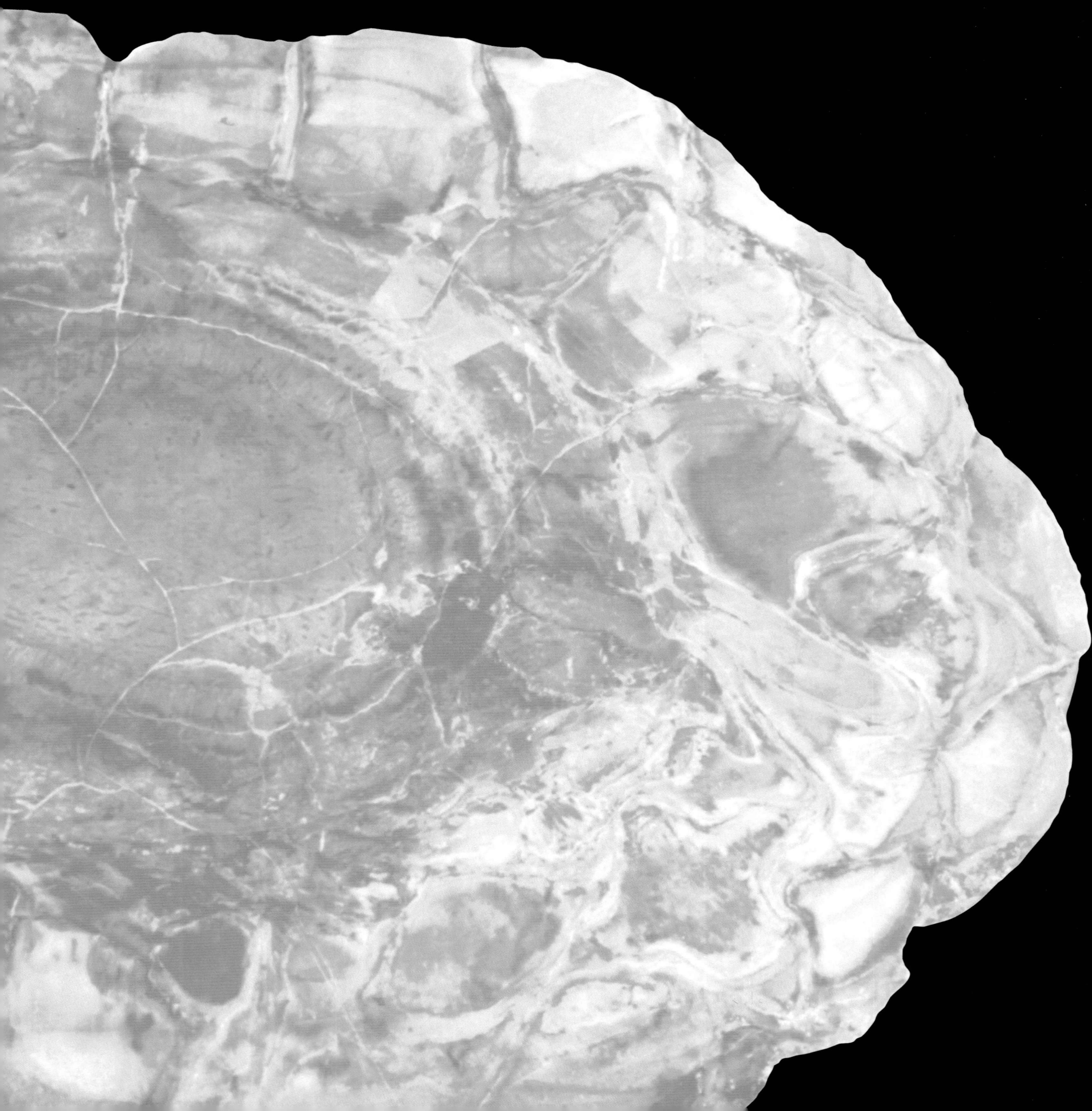

During the Mesozoic, the ferns and conifers, particularly the conifers, had increased in number and importance following their beginning in the Palaeozoic. The "cycadophytes" had also expanded from their origin during the Late Palaeozoic to become an important group during the Mesozoic. In fact, at one time, this era had been termed the "Age of Cycads." The first definite petrified trunks of this group are known from the Triassic.

"Cycadophytes" includes the Cycadales, plants that exist as both living and fossil, and the much larger number of fossil (extinct) types assigned to the Bennettitales (Cycadeoidales), that are commonly referred to as the "fossil cycads."

In many ways these two orders are similar, particularly in respect to their general growth habit and their mostly pinnately divided leaves. Their trunks range from short and squat to tall and columnar and are covered with a protective layer of persistent leaf bases. Cones of the Cycadales occur at the top of the stem (Fig. 1), but are embedded among the leaf bases of *Cycadeoidea* (*Bennettites*) and *Monanthesia* (Figs. 11, 20) and are solitary in *Williamsonia* and other genera of the Bennettitales. The leaf traces pass directly to the leaves in the Bennettitales, but in the Cycadales, they arise from the vascular cylinder at a lower level than the leaf and then spiral around the stem upwardly before finally entering the leaf. These are known as girdling leaf traces.

1 *Encephalartos,* a living cycad showing cones at it's apex

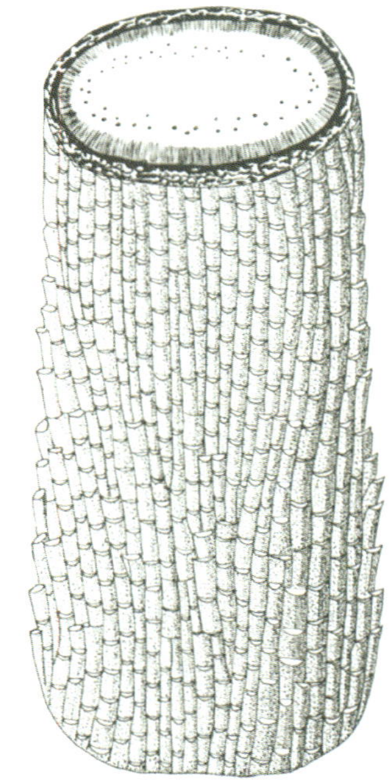

3 Stem of *Lyssoxylon grigsbyii*
Xylem (solid lines), pith (white) with mucilage canals (dots)
0.5 x

2 *Lyssoxylon grigsbyii* (xs)
Arizona and New Mexico, U.S.A.
original size

Cycadales

Fossil stems of true cycads are rare, with those being petrified or permineralized stems being particularly rare. *Antarcticycas*, a permineralized stem from the Early Middle Triassic of Antarctica, may be one. Other Triassic permineralized stems considered to be cycadalean are *Lyssoxylon* and *Charmorgia* from the Chinle Formation of New Mexico and Arizona and *Michelilloa* from the Ischigualasto Formation of San Juan province in Northern Argentina. *Fascisvarioxylon* and *Sewardioxylon* are represented by structurally preserved axes from the Jurassic of the Rajmahal Hills of India, and two Tertiary fossil stems illustrating the cycadalean habit are *Menucoa* from the Rio Negro Province and *Bororoa* from the Chubut Province of Argentina.

Lyssoxylon

The columnar stem consists of a large pith surrounded by the vascular cylinder with the xylem (wood) containing a few growth rings (Fig. 2). The stem is surrounded by persistent, close, oval, steeply inclined leaf bases (Fig. 3). Epidermal hairs form a dense mat around the petiole bases. The leaf traces appear to be girdling.

Charmorgia

The short, stout stem is about 40 cm high by up to 30 cm wide with persistent, elliptical to rhombic-shaped leaf and cataphyll bases (Figs. 4-5). Cataphylls are generally concerned with protection of various parts of the plant. The stem has a large pith, narrow vascular ring, and a broad cortex that appears to have girdling leaf traces (Fig. 6). A number of leaf traces enter each leaf (Fig. 7).

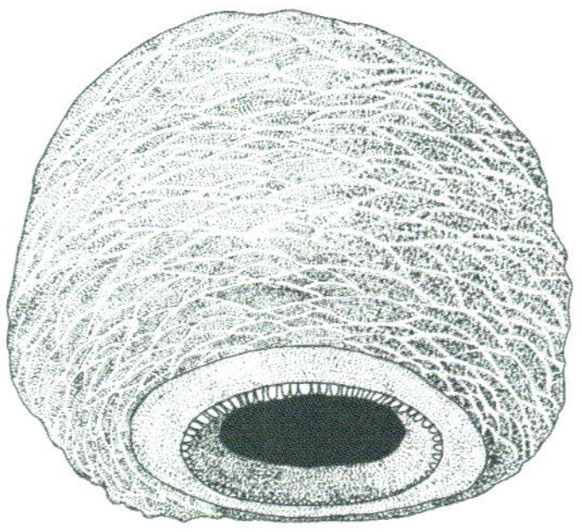

4 Stem of *Charmorgia dijolii*
Pith (black)
Xylem (fine dotted area)
Arizona, U.S.A., 0.25 x

5 *Charmorgia dijolii*
Arizona, 0.5 x

6 Basal view of *Charmorgia dijolii*, 0.5 x

7 *Charmorgia dijolii,* cataphyll or leaf base, 3 x

Michelilloa
The stem is incompletely preserved and about 10 cm in diameter. It has a massive pith containing scattered mucilage canals and has long, filamentous epidermal hairs on its stem surface. The vascular cylinder is divided by leaf gaps into segments. The specimen was collected in the Ischigualasto desert from the same bed as *Rhexoxylon*, the gymnosperm with wedge-shaped xylem (Tidwell, in this volume).

Menucoa and *Bororoa*
Menucoa and *Bororoa* are similar in having persistent leaf bases, numerous mucilage canals and broad rays in their xylem.

Bennettitales (Cycadeoidales)

Williamsonia
This genus has been reported from the Upper Triassic and Lower Cretaceous, but is most abundant in the Jurassic. The name *Williamsonia* was originally given to the whole plant, but now generally restricted to solitary female cones that have been traditionally referred to as "flowers," although they are not like the flowers in angiosperms (flowering plants). The cones consist of a receptacle bearing abundant scale-like leaves (bracts) below (Fig. 9) and numerous, elongated, sterile interseminal scales surrounding very small ovules or seeds above. The ovules or seeds occur on elongated stalks.

The most completely known plant of *Williamsonia* is *W. sewardiana* Sahni 1932 from the Jurassic of India. Its reconstruction consists of a small, columnar tree bearing a crown of leaves (Fig. 8). Small scars resulting from pointed scales occur among the larger leaf scars that form the protective layer around the outside of the stem. Branches on which cones and leaves grew occur at intervals along the stem. *Williamsonia* has been recorded from Great Britain, Mexico, the United States, Canada, Australia, and India.

8 Reconstruction
of *Williamsonia sewardiana*
India, modified from Sahni 1932

9 Cone of ?*Williamsonia* sp. with bract impressions on the outside
Morrison Formation, Utah, U.S.A.
2 x

10 The first *Bucklandia* sp. from the Morrison Formation, Utah, U.S.A., 0.5 x

11 Stem of *Cycadeoidea* sp. showing cones among the leaf bases South Dakota, U.S.A.

12 *Cycadeoidea* sp. illustrating an embedded cone among the leaf bases Argentina, original size

Bucklandia

Bucklandia stems are slender and usually represented by casts and compressions that have often been related to *Williamsonia*, although they may be either cycadalean or bennettitalean. When petrified, the stems have a covering of persistent, often rhomboidal-shaped, large leaf bases (Fig. 10). Where preserved, the large pith is surrounded by a wide secondary xylem and a narrow sequence of vascular cambium, secondary phloem, cortex, and ramentum. The stems may also contain numerous mucilage canals. *Bucklandia* species occur in Great Britain, United States, India, Japan, and very likely will be eventually discovered in Argentina.

13 *Cycadeoidea* sp.
Note the cones (arrows)
Utah, 0.5 x

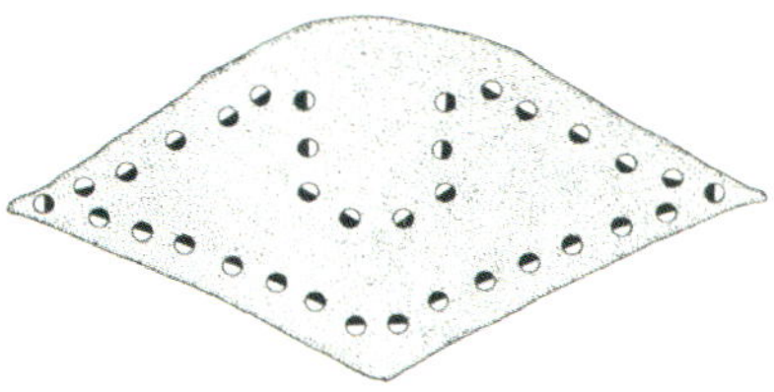

14 A *Cycadeoidea* leaf base (xs) showing the orientation of the leaf traces into the leaf

13a Detail of an embedded cone
Note the row of seeds (arrows)

Cycadeoidea (*Bennettites*)

Most specimens of *Cycadeoidea* are represented by squat trunks, although some may be columnar, having a dense covering of leaf bases containing embedded cones (Figs. 11-13). Specimens range from 15 to 900 mm or more in height. Evidence that cycadeoids have been of interest since ancient times is furnished by the discovery in 1867 of a silicified trunk among vases and other objects of ancient worship in a tomb in northern Italy where it had been placed more than 4000 years before by the Etruscans.

The genus *Cycadeoidea* has been a source of interest and controversy since its discovery on the Isles of Wight and Portland of Great Britain in the early nineteenth century. The first bennettitalean specimens discovered in Great Britain were sandstone casts of *Bucklandia*. Later the name *Cycadeoidea* was given for petrified trunks from the Isle of Portland. Similar trunks were subsequently found on the Isle of Wight and termed *Bennettites*. In the interim, another trunk from near Lednica, Poland, now the type of *Cycadeoidea reichenbachiana*, was given yet another generic name, *Raumeria*. Eventually, it was discovered that there was no appreciable differences between them, and therefore, because *Cycadeoidea* is the older name, it is considered the valid name for the genus. *Bennettites* has still been used for some species from the Isle of Wight, and since the names have often been used interchangeably, has added to

this nomenclatural confusion.
A cross section of a cycadeoid trunk shows a large pith in the center surrounded by a zone of xylem interrupted by large, ray-like structures (Fig. 15). The cortex is parenchymatous and contains mucilage canals and leaf traces. Each leaf trace arises as a single strand but divides into several, before becoming arranged in a "horseshoe" curve upon entering the petiole of the leaf (Fig. 14). Immature leaves have been found near the apex of these trunks, but the size of their mature leaves is unknown. However, the fronds of *C. ingens* have been estimated to have been at least 300 mm long.
The outer protective layer, the ramentum ("armour"), consists mostly of persistent leaf bases that vary from triangular to rhomboidal in shape, but are often distorted (Fig.12). Associated with the bases are long ramental, fusiform-shaped scales that form a rather thick wall around each base. Cones are borne in the axils of leaves and are scattered and nearly completely sunken in the ramentum surrounding the stem (Fig. 13).
The cones are composed of bracts arranged concentrically on an axis, stalk, or peduncle that terminates into a conical or dome-shaped receptacle containing numerous, very small seeds (Fig. 13a). The cones are generally considered to be both male and female with the male portion being crushed as the cone matured and elongated.
The seeds occur between the sterile interseminal scales like in *William-*

15 *Cycadeoidea* sp. (xs) with large pith and narrow xylem (arrow), Utah, 0.5 x

16 Yellowcat area of Utah where many cycadeoids have been collected

17 Collecting fossil "cycads" in the Morrison Formation, Utah

sonia. The tips of the scales are club-shaped which fit into the spaces between the seeds and fuse to form a continuous layer around the cone with openings through which the micropylar beaks of the ovules or seeds project. The micropylar beaks served as passageways through which the pollen grains had moved in order for fertilization to eventually take place. The fused scales protected the seeds until ripe.

Cycadeoidea have been found in Great Britain, France, Belgium, Austria, Germany, Poland, northern Italy, Patagonia (Argentina), Queensland (Australia), India, northern Japan, British Columbia (Canada); an impression of one from Prince Edward Island, Canada; and over a thousand have been collected from the western United States, particularly South Dakota and Wyoming. Many specimens also occur in Nebraska, Montana, Utah (Figs. 16-17), California, Colorado, New Mexico, Oklahoma, Kansas, Texas, North Carolina, and Maryland.

Monanthesia

Specimens of *Monanthesia* consist of squat to mostly columnar stems that are upwards of 2.2 m high and where complete, having a diameter of 51 cm or more across (Fig. 19). In transverse section, the structure of the stems are like the cycadeoids (Fig. 18). The trunks have spiral, persistent, angular leaf bases with only a female cone in the axil of nearly every leaf, apparently all are at the same stage of development. The male plant is

18 *Monanthesia* sp. in transverse section Utah, 0.5 x

20 Cones of *Monanthesia* sp. (ls)
Note the rows of seeds (arrows) in the cones
South Dakota, 1.4 x

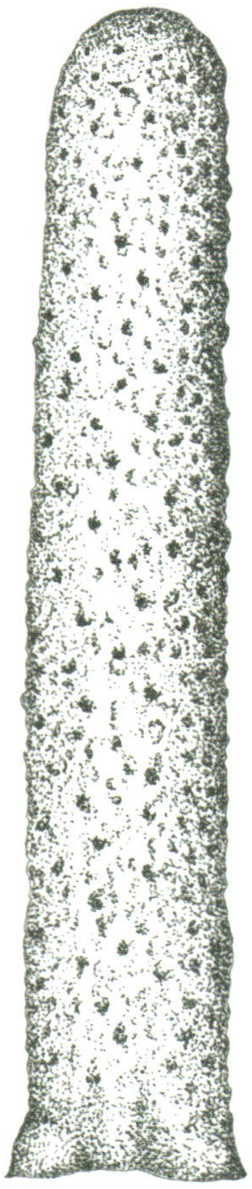

19 *Monanthesia* sp.
Utah, U.S.A., 0.05 x

21 Extended stalks of *Monanthesia* sp.
The lengthening of the stalks pushed the cones beyond the ramentum.
Utah, 0.5 x

presently unknown. The cones are similar to those of *Cycadeoidea* (Fig. 20), however, in some specimens, it appears the peduncle or stalk elongated, extending the cones beyond the ramentum for easier cross-pollination and/or seed dispersal (Fig. 21).

Specimens of *Monanthesia* have been collected from the Early and possibly Late Cretaceous of San Juan Basin (New Mexico), near Moab (Utah), northern California, the Black Hills of South Dakota, Niabrara Chalk Hills of Trego County (Kansas), northeastern Colorado, similar specimens in Maryland (*Cycadeoidea fisherae*), northern Italy, and possibly *Cycadeoidea saxbyana* (Br.) Morris from the Isle of Wight also belongs to this genus.

WILLIAM D. TIDWELL

Hermanophyton - eine rätselhafte Pflanze aus dem Jura

Hermanophyton - an Enigmatic Plant From the Jurassic

Stems of *Hermanophyton* (Hermie wood) are known from only four localities in the Upper Jurassic Morrison Formation in Utah and Colorado of the United States. Stems of this genus range from 30 to more than 225 mm in diameter and up to 18 m or more in length. In cross section, they consist of a relatively narrow, parenchymatous pith, a xylem cylinder composed of 9-15 large, wedge-shaped segments separated by broad primary rays, a two-layered cortex, surrounded by a ramentum of long hairs and small club-shaped appendages (Figs.1-3, 5). Small, round, persistent leaf bases are embedded in the ramentum (Fig. 6).
Hermanophyton is generally seen as a single stem, but one specimen has been collected that branches (Fig. 7). This branching occurs well over 15 m from the plant's base.

Arnold compared *Hermanophyton* to the very similar Southern Hemispheric Triassic form, *Rhexoxylon*. The wedges of secondary xylem in the stems of *Rhexoxylon* occur outside of a primary ring of vascular bundles separated by broad rays (Fig. 8). In the species, *R. pianitzki*, secondary xylem develops on both sides of the primary xylem. Some develops towards the pith and some towards the stem exterior. The large pith contains thin-walled parenchyma, sclerotic nests, abundant secondary parenchyma organized into irregular masses, and anomalous secondary growth. Rhomboidal-shaped leaf bases occur on the outside of the stem. These features are very different from those in *Hermanophyton*. *Rhexoxylon* has been reported from South America, South Africa, and Antarctica.

3 *Hermanophyton taylorii* (xs)
Colorado, Dernbach collection
original size

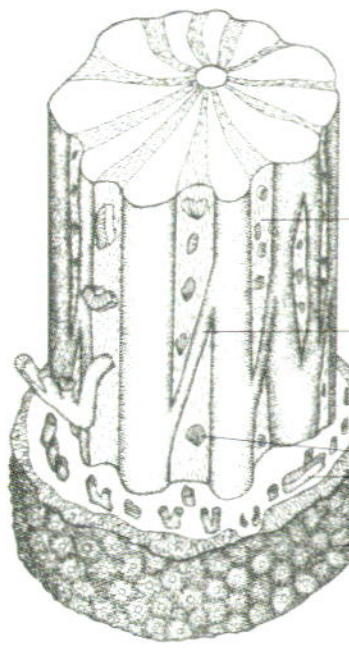

1 Reconstruction of
Hermanophyton glismannii
Colorado, U.S.A., 0.3 x

4 *Hermanophyton glismannii* (xs)
Colorado, original size

2 *Hermanophyton taylorii* (xs)
Colorado and Utah, original size

5 *Hermanophyton ownsii* (xs)
Colorado, original size

6 *Hermanophyton taylorii* (ls)
Outer surface, note the weathered, round leaf bases
Colorado, original size

7 Branching of *Hermanophyton*
Portion of specimen is missing

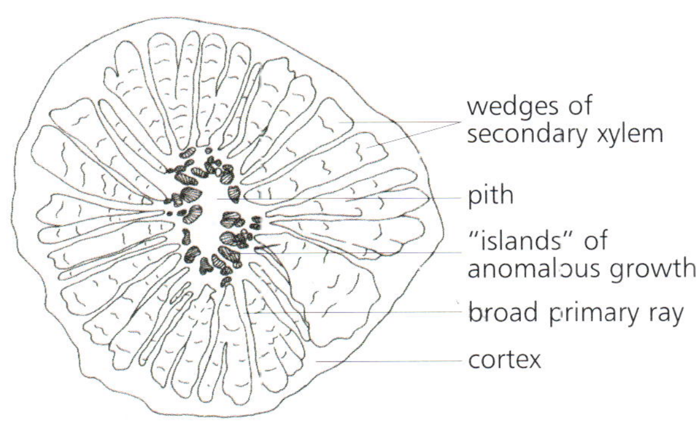

8 *Rhexoxylon tetrapteroides* (xs), South Africa
idealized reconstruction after Walton 1923

WILLIAM D. TIDWELL

Heutige Baumfarne und ihre fossilen Verwandten

Extant Tree Ferns and Their Fossil Relatives

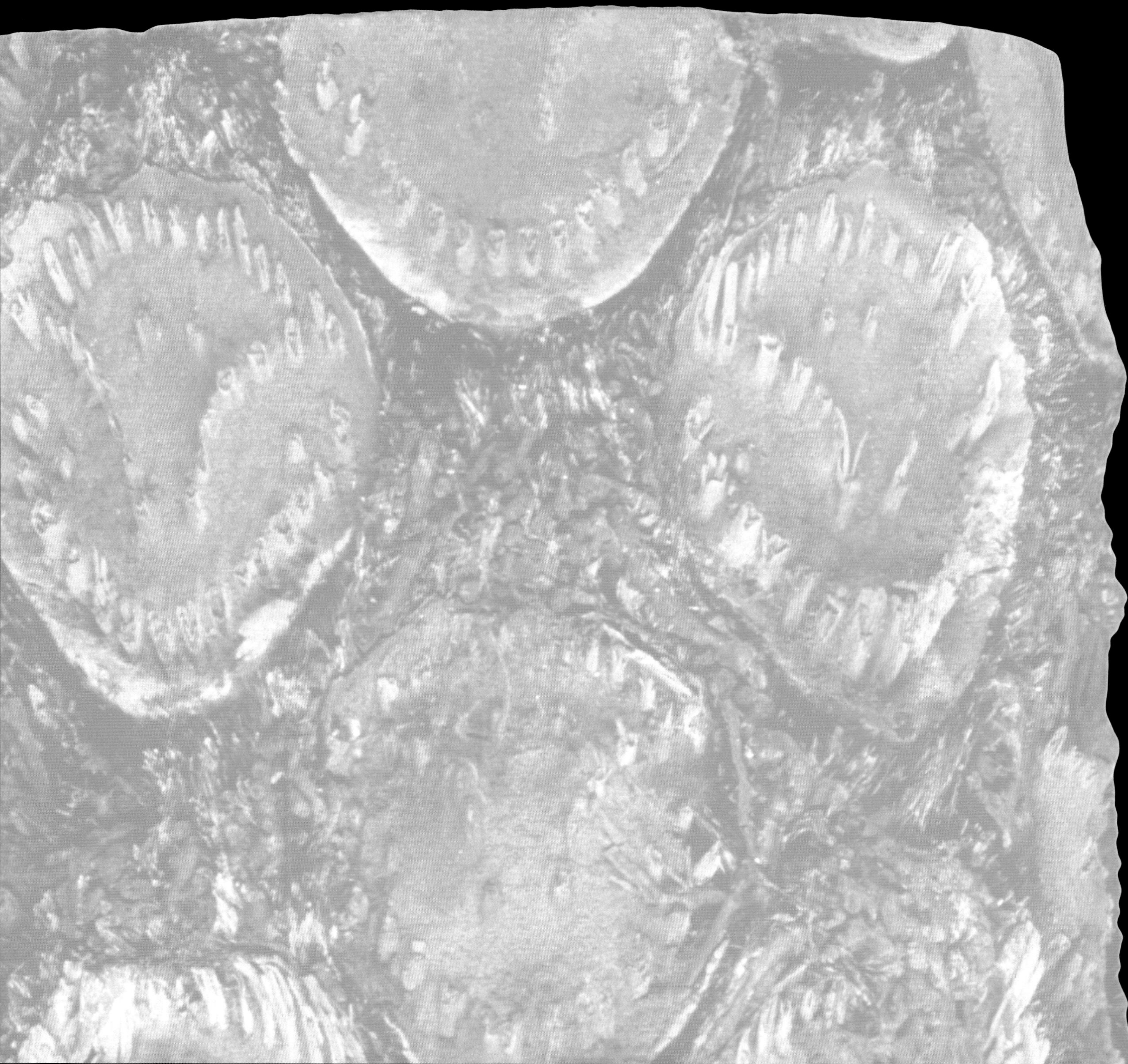

Tree ferns were once an important part of the world's flora, and remain a conspicuous part of the world's tropical rainforests today. Approximately 680 species of ferns generally exhibit some form of the tree fern system of growth. Tree ferns are usually placed in the family Cyatheaceae or Dicksoniaceae with the Cyatheaceae embracing the majority of the living species. Some researchers, however, consider them to be one family and assign them all to the Cyatheaceae, whereas others have subdivided them further, and, thus, propose many families.

Tree ferns usually consist of tall trunks having large, complex pinnate leaves that form a crown at their apices (Fig. 1). The trunk is usually massive and stout (Fig. 2). In the older plants, a thick mat or mantle of roots, arising directly from the stem and known as adventitious roots, compose the main anchorage and support for the stem. Depending upon the genus, the vascular cylinder of the stem varies from a slightly dissected siphonostele (Fig. 8) to where the vascularization is divi-ded into segments forming a true dictyostele (Fig. 3). In the forms having a dictyostele, the individual vascular strands or meristeles may or may not be V-shaped. Two of the main anatomical differences between the two families are the medullary bundles that generally occur in the pith of the cyatheacous stems, but not in the dicksoniaceaous types, and the sclerenchymatous band around the main vascularization in the Cyatheaceae, but is lacking in the Dicksoniaceae. They both appear to have been in existence since, at least, the middle Mesozoic era.

Nishidacaulis

The stem of *Nishidacaulis burgii* appears to have radial symmetry and is surrounded by persistent leaf bases and adventitious roots (Fig. 4). The stem is solenostelic, that is phloem surrounds it's xylem. It has undivided leaf traces that are modified reverse omega- or U-shaped having incurved ends with lateral constrictions on each side. Multicellular epidermal hairs and mucilage cells are also present. This species was collected from an Eocene conglomerate in northwestern Nebraska that was most likely reworked from sediments of Early Cretaceous age in the Black Hills, South Dakota, U.S.A. Species of *Aurealcaulis*, *Tempskya*, and *Cycadeoidea* have been collected associated with it.

Conantiopteris

The stem of *Conantiopteris* is dictyostelic (Fig. 5) with its outer surface covered by persistent leaf bases, adventitious roots, and multicellular trichomes or epidermal hairs. The oval-shaped leaf bases show leaf traces divided into three parts. The

1 *Cyathea* sp. tree fern
New Zealand

2 *Dicksonia* sp. tree fern
Victoria, Australia

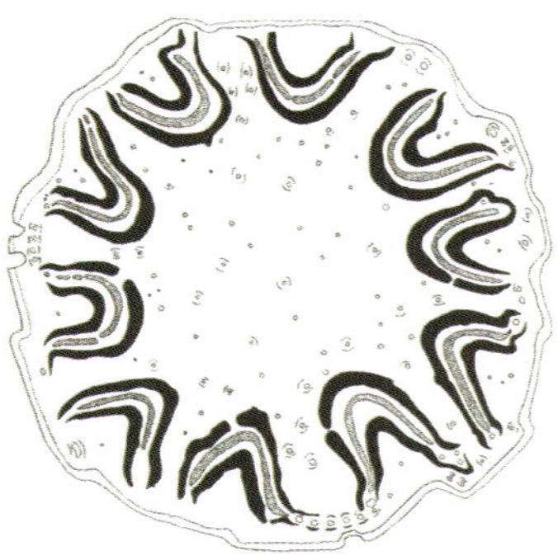

3 A dictyostelic stem of *Cyathea* (xs) showing V-shaped meristeles and leaf traces; vascularization (stipples), sclerenchymatous bands (black), and pith with medullary bundles (dots) in the middle
0.5 x, modified from Ogura

4 *Nishidacaulis burgii* (xs)
Note the reverse omega- or U-shaped leaf traces
Nebraska, U.S.A., original size

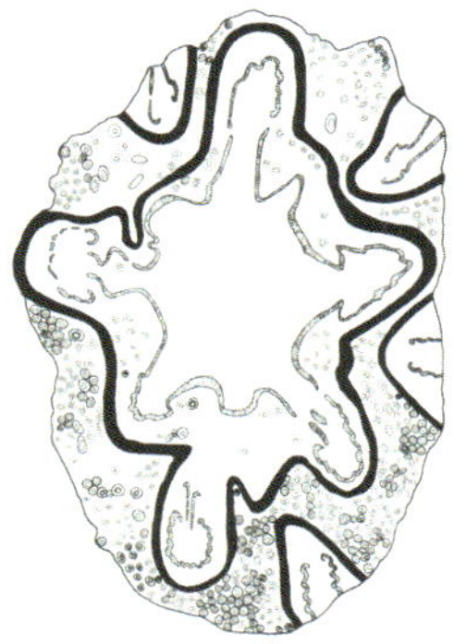

5 Stem of *Conantiopteris schuchunenii* showing general features of anatomy, pattern of leaf divergence, and numerous adventitious roots (xs)
California, U.S.A., 0.5 x
modified from Lantz et al. 1999

8 *Cibotium iwatense* (xs)
Vascularization (stippled), hypodermal sheath (black)
Japan, original size
modified from Ogura 1933

6 "*Cibotium*" *tasmanense* (xs)
Lune River, Tasmania, Australia
original size

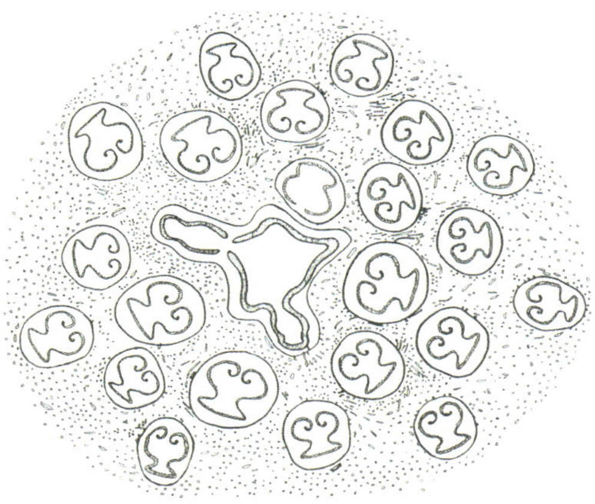

7 "*Cibotium*" *tasmanense* (xs)
Stem with surrounding petioles and adventitious roots, 0.4 x

two layered cortex of the stem has a parenchymatous inner cortex and a sclerenchymatous outer. No medullary or cortical bundles are present. The species was collected from the Lower Cretaceous Budden Canyon Formation at the Shoup Ranch Gulch in northern California, U.S.A.

Cibotium

"*Cibotium*" *tasmanense*

This species is based on a stem with a mantle of persistent leaf bases, adventitious roots (Fig. 6), and long, filamentous epidermal hairs. The vascularization of the stem is convolute due to the development and departure of the leaf traces (Fig. 7). No sclerenchymatous sheath is present around the vascular strands. The U-shaped leaf traces depart the stem vascularization (stele) leaving a small leaf gap. The ends of leaf trace nearest the stem incurve after separation (Fig. 7). The flattened sides of the trace form slight lateral grooves or indentations.

"*Cibotium*" *tasmanense* should be placed in a genus other than *Cibotium*, possibly something new. "*Cibotium*" *tasmanense* was originally considered to be Tertiary from the Lune River locality, but later demonstrated to be mid-Mesozoic in age.

Cibotium iwatense

A small, solenostelic stem, (xylem surrounded by phloem) with spirally arranged leaf bases constitutes this species. The vascularization of the stem is nearly continuous, interrupted only by leaf gaps (Fig. 8). A hypodermal sheath is present. The species was reported from the Upper Cretaceous near Utsume, Kunohe-Gun, Iwate-Ken, Japan.

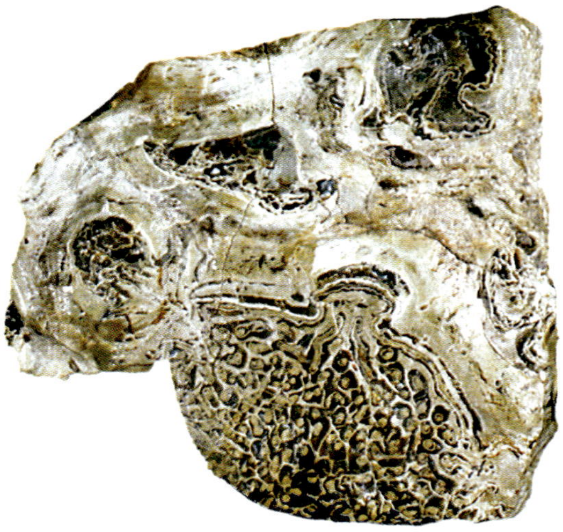

12 *Oguracaulis banksii* (xs)
Note the departing leaf trace and medullary bundles in pith, original size

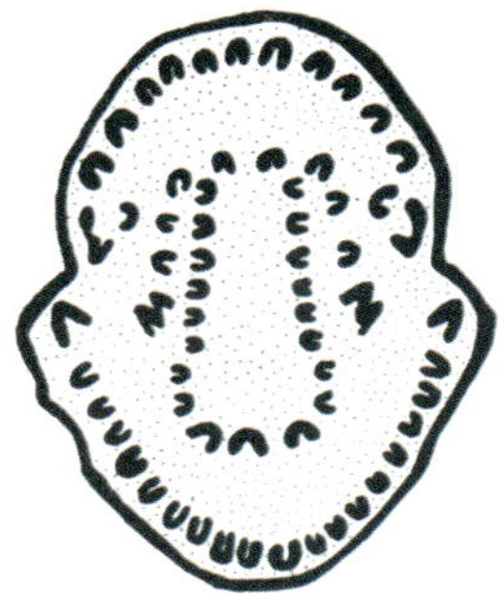

9 *Cibotium oregonensis*
petiole, Oregon, U.S.A
original size

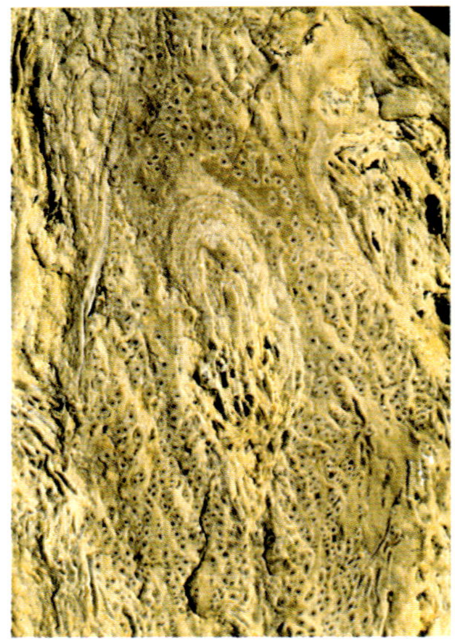

11 *Oguracaulis banksii*
Side view of specimen, note leaf bases and roots on surface
Lune River, Tasmania, Australia
original size

10 *Protopteris* sp.
Part of stem with leaf bases showing their vascular strands
0.3 x, modified from Corda 1845

Cibotium oregonensis

This species was defined from leaf bases and a portion of its stem, both of which are covered by multicellular epidermal hairs. The vascularization of the stem lacks a sclerenchymatous sheath and its petioles are divided into a number of vascular strands (Fig. 9). Adventitious roots arise from the stem. The species was described from material collected from Eocene strata near Medford, Oregon, U.S.A.

Protopteris

Protopteris is similar to modern tree ferns. Persistent leaf bases and numerous adventitious roots cover the stem surfaces (Fig. 10) and its vascular system is similar to living *Cyathea*. The margins of the leaf gaps, however, turn deeply towards the pith where they appear to connect with medullary bundles. This type is also found in *Cyathocaulis*.

The genus has been reported from the Upper Greensands beds in Dorsetshire, England and the Lower Cretaceous of Bohemia, Czech Republic.

Oguracaulis

Oguracaulis banksii consists of specimens from Upper Jurassic-Lower Cretaceous strata at the Lune River locality near Ida Bay, Tasmania, Australia. The specimens are large fern stems with associated leaf bases surrounded by a thick root mantle (Fig. 11). The stems are dictystelic with a sclerenchymatous sheath, numerous medullary bundles in the pith, undivided, omega-shaped leaf traces, and multicellular epidermal hairs (Fig. 12).

Cyathocaulis

This genus is assigned to stems resembling living *Cyathea*. The stems are dictystelic with each of the meristeles surrounded by a sclerenchymatous sheath (Fig. 13). It has numerous adventitious roots and multicellular hairs on the outside. The pith contains medullary bundles and occasional mucilage canals and root traces. The vascularization at the lower part of the leaf base is V-shaped. *Cyathocaulis* differ from *Cyathea* by having inwardly curved stelar margins of the meristeles at the leaf gaps and having medullary root traces (Fig. 13). *Cyathocaulis* species have been reported from the Jurassic of N. Kyöngsang Dô, South Korea and the Cretaceous of Wakayama Perfecture and Choshi, Japan, as well as from the Sumagui River, Bangabon, Mindoro, Phillipines.

Cibotiocaulis is another Cretaceous genus that shows many similar anatomical characters to *Cyathocaulis*, but differs in having smaller raised leaf scars and a thinner root mantle.

Alsophilocaulis

Alsophilocaulis is presently represented by only a single silicified specimen. The stem is dictystelic with V- and W-shaped strands similar to *Cyathea* (Fig. 3). An abundance of sclerenchyma is associated with the xylem. Its pith contains numerous tracheids and fibers. *Alsophilocaulis calveloii* is from Tertiary sediments in Arroyo Puipucón, Neuquén/Argentina.

14 *Cyathodendron texanum*
Note the leaf bases.
Texas, U.S.A., original size

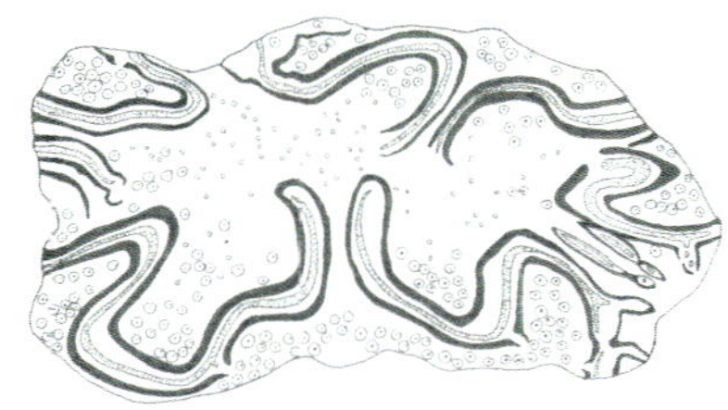

13 *Cyathocaulis yabei* (xs) showing the inwardly curved ends of the leaf gaps
vascularization (stipple)
sclerenchymatous sheath (black)
medullary bundles (small circles)
roots (circles with dots)
Mindoro, Philippines, original size
modified from Ogura 1941a

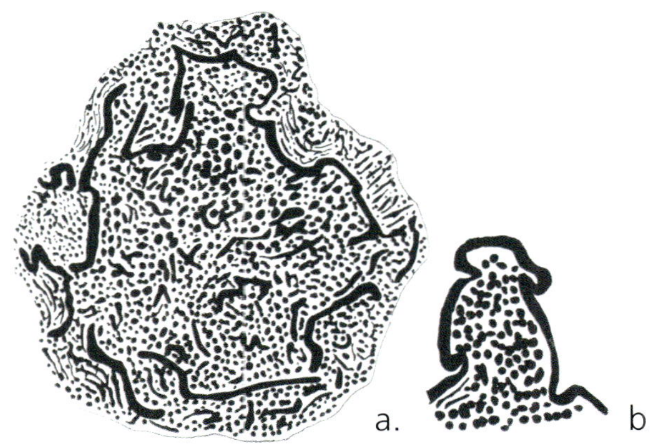

15 A stem of *Cyathodendron texanum* (xs) with numerous medullary bundles (a; x 1) and departing leaf trace (b)

Dendropteridium

The stem of this genus shows the typical cyathean dictystelic structure (Fig. 3). There is a single series of peripheral concentric meristeles which turn out at the margins of the leaf gaps. Numerous small medullary bundles occur irregularly throughout the pith and are related to the formation of the leaf traces. The leaf traces are formed of separate bundles as well.

The species, *D. cyatheoids*, was reported from a Late Tertiary volcanic agglomerate near Butlandiga on Mount Elgon, Kenya.

Cyathodendron

Cyathodendron texanum is a cyatheaceous-like fern from the Eocene of Texas. The stems of *C. texanum* are upright and covered by a mantle of spirally arranged leaf bases, adventitious roots, and multicellular epidermal hairs (Fig. 14). The vascular system is an irregularly shaped solenostele with numerous medullary bundles in its pith (Fig. 15). Leaf gaps are present but appear to be closed by a weird downward extension of the vascular tissue, leaving an oblique opening through the vascularization of the stem. The leaf trace near the base of the leaf is continuous and trough-shaped with incurved margins, but eventually divides into separate vascular strands at higher levels of the leaf.

Incertae sedis

Yulebacaulis

Silicified material of *Yulebacaulis normanii* consisting of outer cortex, petioles, small vascularized appendages known as aphlebiae, and roots (Fig. 16) were collected from near the Combabula Station north of Yuleba, southeastern Queensland, Australia. The specimens are associated with mid-Mesozoic age osmundaceous ferns. The petioles lack stipular wings and has clepsydroid-shaped leaf traces that are oblong, straight, and elongated tangentially to the stem. Peripheral "loops" occur at the ends of these strands which open as the C-shaped aphlebiae traces depart. The specimens occur as cobles in the Lower Cretaceous Mooga Sandstone (Fig. 17). If *Yulebacaulis* is Early Cretaceous in age, then it could extend the age of the Zygopteridales from Upper Palaeozoic into the mid-Mesozoic.

16 *Yulebacaulis normanii* (xs)
Queensland, Australia, original size

17 Locality for *Yulebacaulis normanii* near Combabula Station, Queensland, Australia

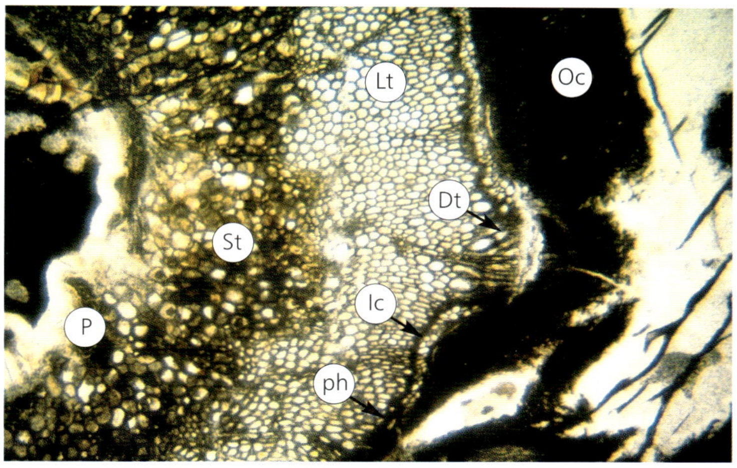

18 A portion of the stem of *Grammatocaulis donponii* (xs) showing beginning of leaf trace
P (pith), St (short tracheids), Lt (long tracheids)
ph (phloem), Ic (inner cortex), Oc (outer cortex)
Dt (departing leaf trace)
Queensland, Australia, 10 x

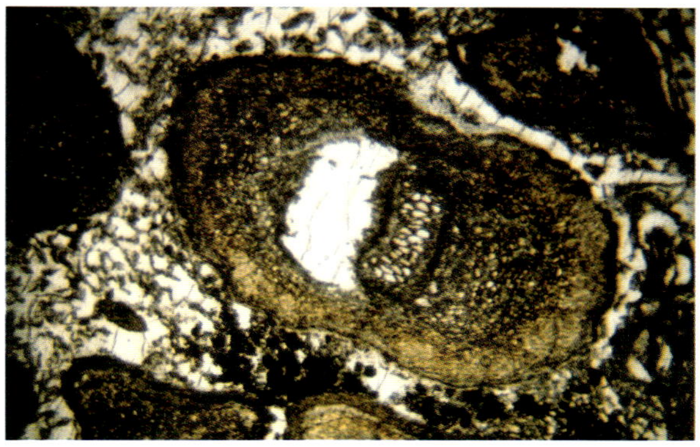

19 A petiolar vascular strand of *Grammatocaulis donponii* (xs), 10 x

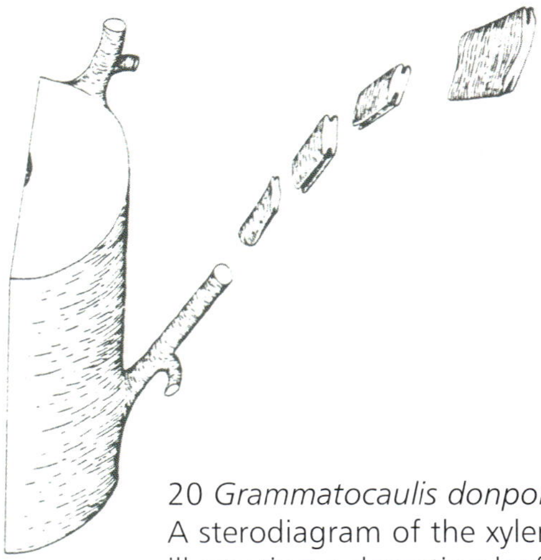

20 *Grammatocaulis donponii*
A sterodiagram of the xylem cylinder illustrating a departing leaf trace originating as round in shape and becoming flattened at higher petiole levels; cortex and outer tissues not shown, not to scale
after Tidwell & Rozefelds 1990

Grammatocaulis

The stem of *Grammatocaulis donponii* is an ectophloic siphonostele, that is phloem only outside of the xylem. The xylem is composed of an outer zone of long tracheids and an inner zone of short, wide tracheids surrounding the pith (Fig. 18). Leaf traces are round in cross section in the inner cortex (Fig. 20), becoming linear to oblong in the outer cortex and outside of the stem. The vascular strands of the petioles are elliptical (Fig. 19). The leaf bases have sclerotic rings, but lack stipular wings like in the Osmundaceae.

This fern stem was collected from the Middle to Late Jurassic Injune Creek Group in the Wandoan area of Queensland, Australia.

Associated with this fern is "pentoxylalean"-like wood, undescribed cycadophytean stems, as well as abundant petrified conifer wood, and osmundaceous axes. Although the family of this genus is presently unknown, it is similar anatomically to *Grammatopteris* of the Anachoropteridaceae from the Late Palaeozoic of Germany (Chemnitz), France (Autun), and now *Grammatopteris*-like specimens from Brazil.

ALFRED SELMEIER

Das nordalpine Molassebecken – eine Schatztruhe für den Sammler versteinerter Hölzer

The Northern Alpine Molasse Basin - A Treasure Trove for Collectors of Petrified Woods

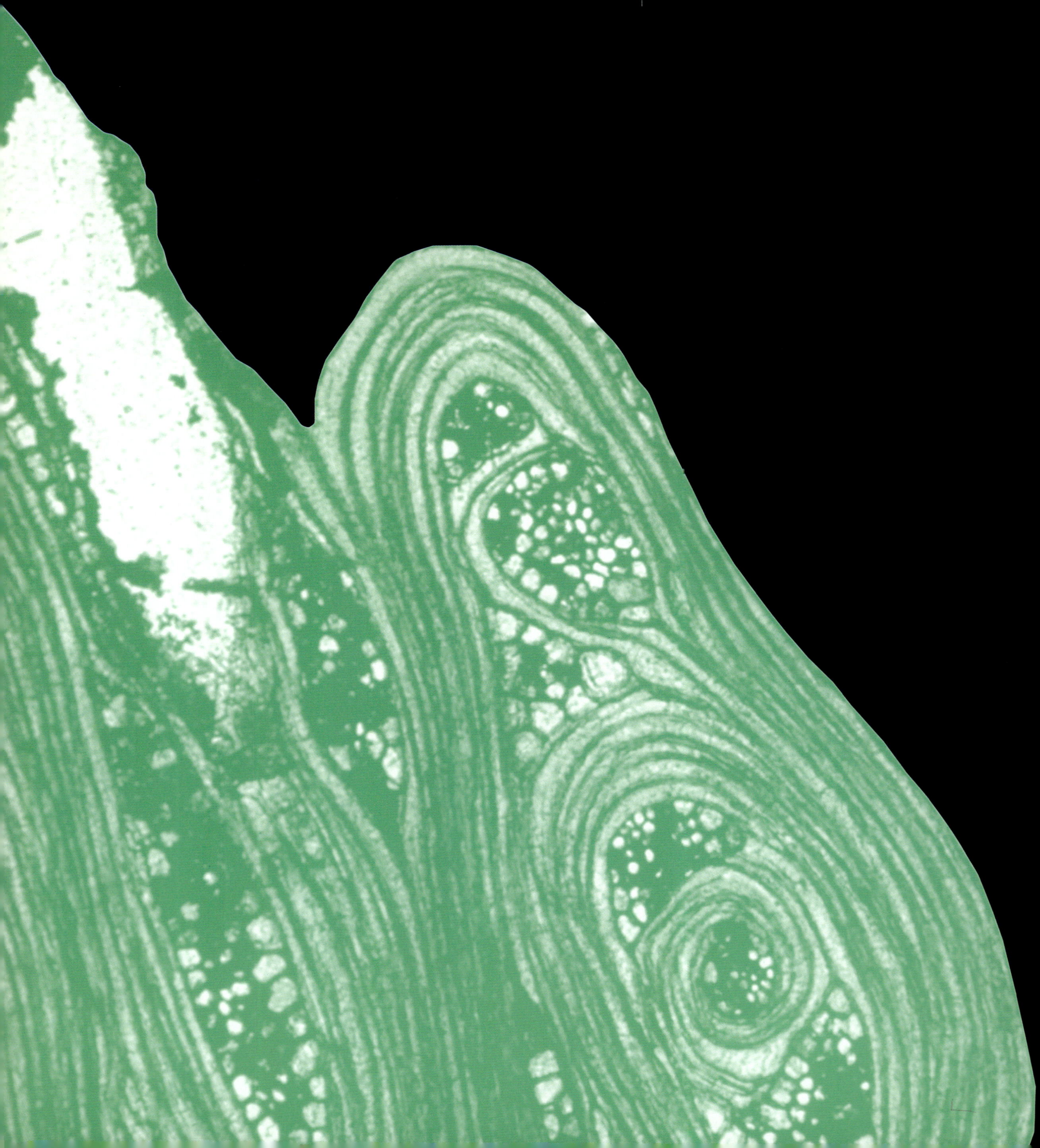

Molasse Troughs, Tertiary Debris Basins

The Molasse deposits stretch across hundreds of kilometers along the northern and southern edges of the Alps. Their depositional areas are east-west oriented debris basins dating from the Tertiary. Their tectonic position and subsequent filling with various sediments are very closely associated with the upward folding of the Alps. The term "molasse" is derived from the French word *molasse* (for "soft") and was originally used for soft, only slightly hardened sandstones found in the region around Geneva. Presently, the word molasse is used to describe either the entire time period during which this sedimentation occurred or to designate the deposited masses of sediments themselves and their geographic distribution. The relatively uncompacted material deposited either as "marine molasse" or as "freshwater molasse." Sands and calcareous marls predominate in the marine deposits. Freshwater molasse, on the other hand, consists typically of alternating layers of sands, marls, and gravel. The large, arc-shaped Molasse Basin in the north, with deposits in Switzerland, southern Germany, and Austria, has its "twin" in the plains around the Po

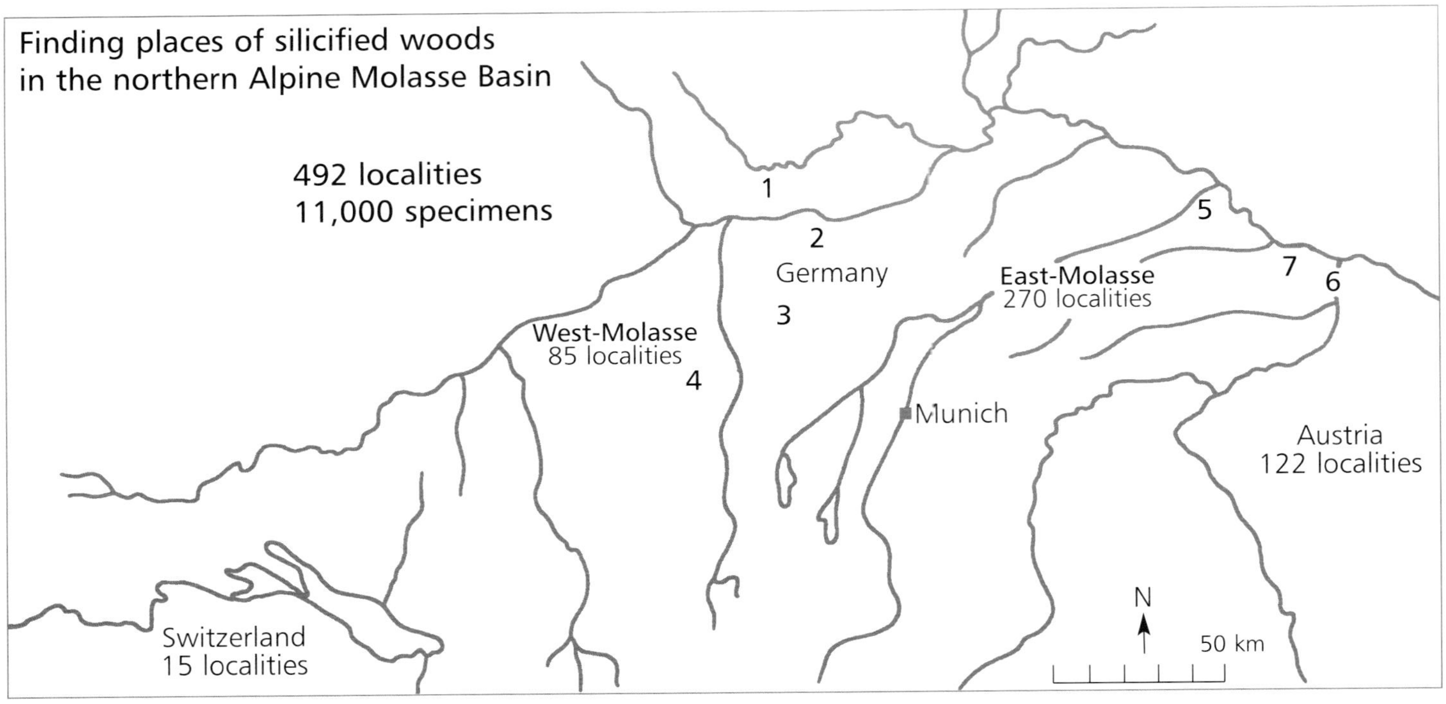

Map of the northern Alpine Molasse Basin with the numbers of finding places in Switzerland (15), southern Germany (85/270) and Austria (122)

River along the southern edge of the Alps in northern Italy. To the west, the northern Alpine Molasse Basin begins in the region of Chambery, SW of Lake Geneva. This basin stretches in a W-E direction for nearly 1,000 km, all the way to the eastern end of the Alps and onwards to the western edge of the Carpathian Mountains. In Bavaria, the Molasse attains an above-ground width of approximately 100 km in a S-N direction. In the north, the molasse strata overly Jurassic limestones; to the east, the Molasse borders on the crystalline basement of the Bohemian mass. To a limited degree, eroded debris from the Franconian Alps and the Bohemian mass was swept into parts of the northern Alpine Molasse Basin.

A Tertiary Window on the Earth's History

The mighty marine or freshwater deposits of the Molasse are a sheer inexhaustible reservoir of fossil documents in the search for Tertiary animal and plant remains. Palaeobotanist W. Jung writes in one publication about a fascinating "window" through which to view the Tertiary era of the Earth's history. The northern Alpine Molasse has yielded not only fossils from numerous animal groups, but also abundant material containing plant fossils such as leaves, seeds, fruits, and pollen. Woods from the Molasse augment the reconstruction of the past vegetation. The collections of museums and private individuals contain thousands of specimens of petrified Molasse woods, usually fist-sized, that were found in ploughed fields. The approximately 18,000 individual specimens counted thus far from the northern Alpine Molasse Basin represent one of the world's most comprehensive collections of silicified Tertiary woods.

Silicified Woods From Switzerland

Compared to southern Germany and Austria, discoveries of petrified woods are rather rare in the Swiss Molasse. A specimen of *Palmoxylon* wood from Lausanne was first described by Stenzel in 1904. Silicified *Laurinoxylon* trunks from strata of the Upper Marine Molasse lay in the river area of the Urnäsch near St. Gallen. Nearly 15 other, geographically distributed, individual finds are documented, including the recent discovery of a cypress trunk in the catchment area of the Bütschelbach in the "Schwarzenburger Land)" Canton Bern (Bern Museum of Natural History). In addition to dispersed and individual finds, Switzerland has thus far yielded only one comprehensive collection of silicified woods. Swiss geologist R. Buxtorf spent more than 90 full days inspecting the Delsberg Basin, where he collected a total of 715 loose specimens on fields. The degree to which they have been rounded, the hardness, and the coloration of many of these woods are similar in appearance to their dark accompanying gravel. In many cases, thin sections must first be cut and prepared before identification can be made with certainty.

Vosges Deposits in the Middle Miocene

The Delsberg Basin, 34 km SW of Basel, is a Tertiary basin surrounded by folded Jurassic strata. The basin's altitude ranges between 415 and 472 meters above sea level. By analyzing the heavy metals and the gravel, it was possible to garner information about the kind of material accompanying the silicified woods. The woods and gravel were transported to and deposited within the Delsberg Basin prior to the upward folding of the surrounding Jurassic strata. The source of the deposited material is the southern Vosges Mountains. It contains unsorted remains of Palaeozoic, Mesozoic, and Cenozoic woods. The Bois de Raube (Robe) Formation of the Delsberg Basin contains the older, Middle Miocene portion of the

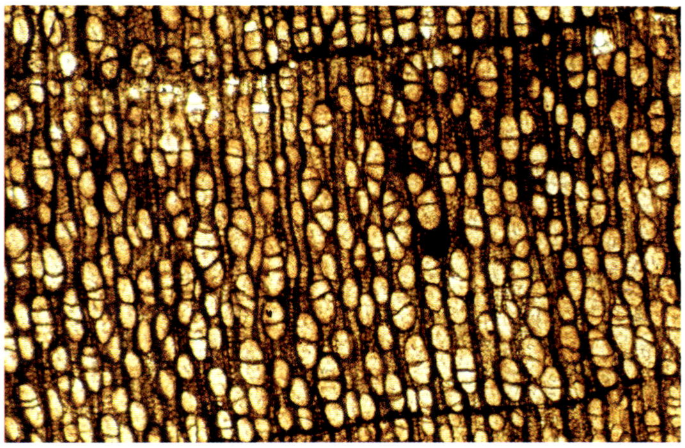

1 *Populoxylon priscum* poplar, Salicaceae family, found at Neuburger Wald, Lower Bavaria. Cross-section with two growth areas; tangential diameter of tracheids in the springwood 75 μm

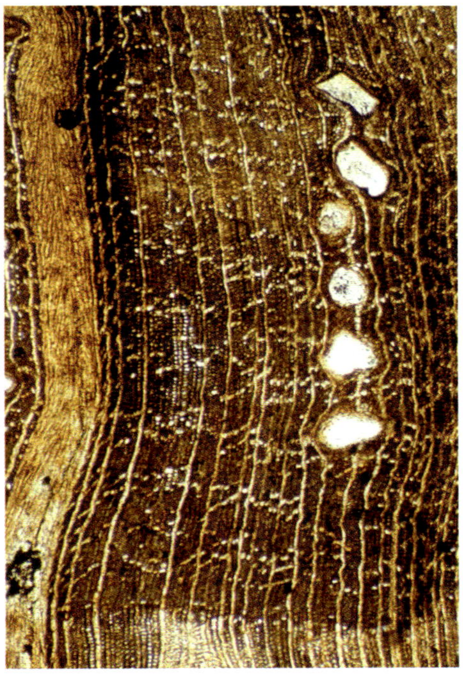

2 *Lithocarpoxylon* sp. oak, Fagaceae family, found at Bad Abbach, Lower Bavaria. Cross-section in the area of the late wood; tangential diameter of tracheids 120 μm

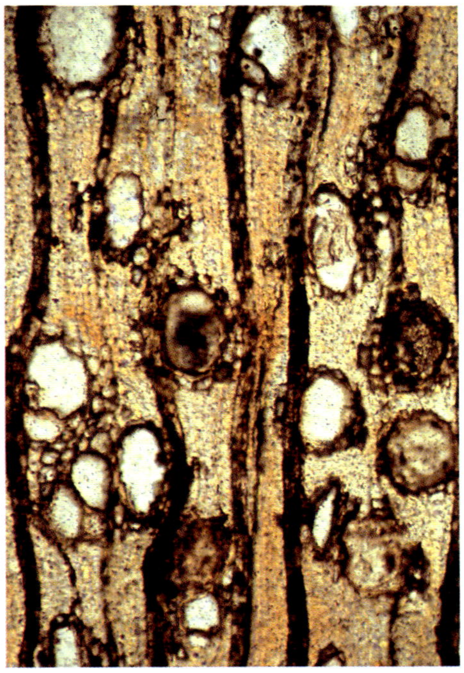

3 *Cinnamomoxylon* cf. *limagnense* cinnamon tree, Lauraceae family. Found at Rauscheröd, Lower Bavaria. Cross-section with paratracheal parenchyma cells; tangential diameter of large tracheids 150 μm

Vosges deposits in the Delsberg Basin. The age of these deposits is believed to be 11-14 million years. The biostratigraphic dating is based on research conducted on two tons of sedimentary material in which Swiss geologist G. Kälin discovered, among other items, the fossil remains of 36 different mammalian taxa. Only the Mesozoic and Palaeozoic woods display a strong superficial similarity to the accompanying gravel. The fluvial transport caused obvious percussive figures in both the woods and the gravel. Characteristic accompanying gravels for these woods are violet Vogesen quartzite, reddish brown flint, porphyries, diorite, marl, and sandstones. Suppositions about the source of these Palaeozoic woods suggest a fluvial transport of debris from the Lower Permian of Faymont or from other regions in the southern Vosges. For example, the microscopic fine structure of a dark specimen of silicified wood from Les Lavoirs looks very similar to that known from the extinct Bennettitales. The identification of this specimen as *Paradoxoxylon leuthardtii* Kräusel is based on 11 thin sections. In phylogenic and taxonomic terms, similarly woods have thus far been reported only from the Keuper discovery site at Neuwelt near Basel and from Thuringia in Germany. The Tertiary woods of the Delsberg Basin are readily recognizable in the field. Their appearance differs from that of the dark and extremely hard accompanying gravel.

5 *Celtixylon cristalliferum* Zürgelbaum, Ulmaceae family, found at Attenfeld, southern Franconian Alb, Bavaria. Cross-section with one growth boundary; tangential diameter of tracheids in the springwood 130 μm

4 *Castanoxylon bavaricum* edible chestnut, Fagaceae family, found at Neuburger Wald, Lower Bavaria. Cross-section with two growth boundaries; tangential diameter of large tracheids in springwood 310 μm

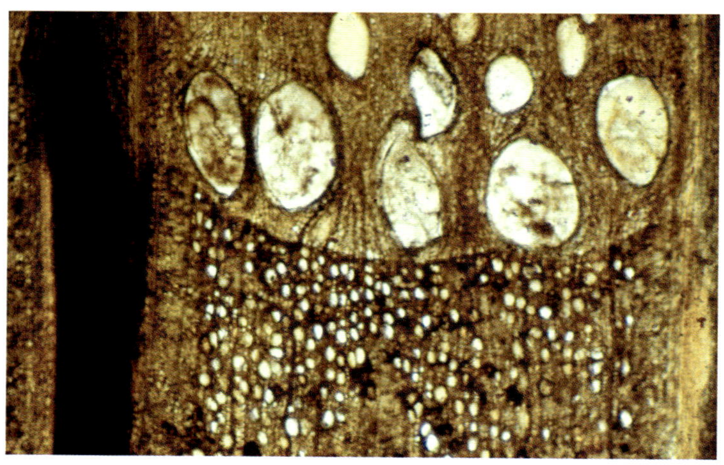

6 *Quercoxylon bavaricum*, oak, Fagaceae family, found at Aidenbach, Lower Bavaria. Cross-section in the area of a growth boundary; tangential diameter of tracheids in the springwood 230 μm

7 *Grewioxylon* sp., Grewia, Tiliaceae family, found at Rauscheröd, Lower Bavaria. Cross-section in the area of a growth boundary; tangential diameter of tracheids in the springwood 180 μm

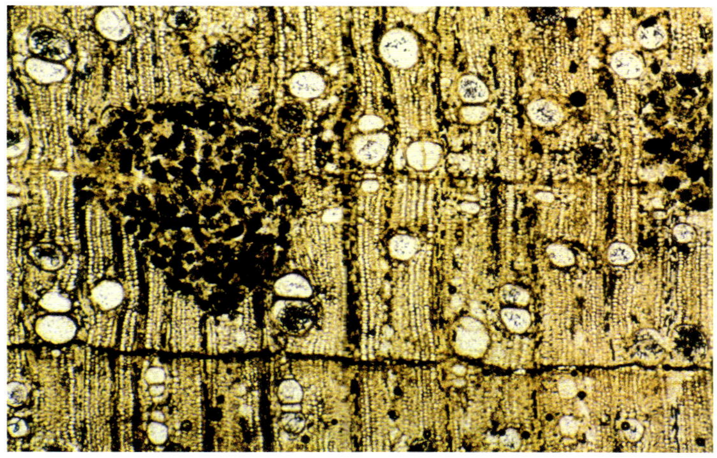

8 *Laurinoxylon* cf. *seemannianum*, laurel, Lauraceae family, found at Egweil, southern Franconian Alb, Bavaria. Cross-section with two feeding tunnels and coprolites of larvae of the *Anobium* sp. beetle. Diameter of large tunnel 800 μm

Quartz crystals, which look somewhat like grains of sugar, encrust their partly rough surfaces. The smallest specimens weigh only a few grams; large specimens can weigh as much as 800 kg. In the majority of cases, the microscopic wood structure is only moderately well preserved. The most frequently identified groups are: Gymnospermae, Fagaceae, Lauraceae, Leguminosae, Juglandaceae, Palmae, as well as a number of as-yet unidentified deciduous woods. A specimen of silicified elm wood was found in a sand pit near Bonfol in Elsgau, Pais d'Ajoie, 9 km NE of Porrentruy. Samples of silicified wood from the Delsberg Basin are kept at the Museum of Natural History in Basel.

Silicified Woods From Austria

Beyond the border defined by the Inn and Salzach rivers, the southern German Molasse continues eastwards into neighboring Austria. The Molasse zone forms Austria's largest contiguous Tertiary area. In the western portion, it primarily encloses the area between Salzach, Innviertel, Linzer Basin, and Enns. The sediments here range in age from the Upper Eocene to the Pliocene. North of the Danube, the Molasse stretches as far as Thaya, which was formerly described as the "outer Alpine Viennese Basin." There are Miocene strata from Eger to Pannon.

Research into the fossil woods of Austria began about 150 years ago, in the mid 19th century. In 1842, F. Unger, a pioneer in the field of Tertiary botany, investigated fossilized trunks and reported on the silicified woods in the National Museum in Linz. In his work on the "Fossil Flora of Gleichenberg", Unger first used microscopic characteristics to determine the identity of wood species.

In subsequent years, the identification of Austrian fossilized woods is primarily associated with names such as E. Hofmann, W. Wurzinger, F. Kümel, W. Berger, B. Kubart, and W. Rössler. The best-known work is the laudable treatment of Eocene phosphorite woods from Prambachkirchen by

E. Hofmann. Using polished surfaces, she reported on 61 different woods, including mangroves. A temporal gap exists in the treatment of Austrian Tertiary woods during the second half of the 20th century. In the 1980's, J.P. Gross of Lyon examined the anatomy of only a few silicified woods from the Waldviertel. The active research of earlier decades didn't resume until 1988, when Cichocki published a study entitled "Zur Histologie tertiärer Hölzer Österreichs". Silicified woods are frequently kept in publicly accessible, state or church collections, e.g. at Museum Eggenburg, Landesmuseum Linz, Kloster Kremsmünster, and Stift Lambach. Austria's Tertiary silicified woods are distributed among 120 different discovery sites. The genera identified thus far are *Glyptostroboxylon*, *Pinuxylon*, *Taxodioxylon*, *Aceroxylon*, *Alnoxylon*, *Liquidambaroxylon*, *Acacioxylon*, *Palaeo-Robinioxylon*, *Platanoxylon*, and *Quercoxylon*. Additional individual finds include *Taxodioxylon* from Trimmelkam, *Castanea* from Bad Zell, *Ilex* (?) from Gallspach, and *Platanus* from Altenmarkt and Pfaffing.

Silicified Woods From Southern Germany

There are four terms which were formerly used to describe petrified woods but which are now only seldom used: "wood-stones," "tree-stones," "dendrolites," and "Nicolias." A glance at the geological map of Bavaria (drawn to a scale of 1:50,000) offers general directions in the search for "wood-stones." The yellow areas on the map indicate the location of the Upper Freshwater Molasse and thus provide an initial clue to sites where it might be possible to find specimens. The following localities have often yielded remains of petrified wood: pebbles from the Danube, gravel, sand, and clay pits in undulating, hilly, Tertiary land; exposures of the "Ortenburger Schotter" with the so-called "tree-holes"; brook gorges in Neuburger Wald near Passau. Tertiary conglomerates in glacial-fluvially covered regions of the Alpine foothills; the gravel in Alpine bodies of water; depositions of gravel dating from the Ice Age; loose stones found on cultivated fields, e.g. harvested grain and sugar beet fields in the southern Franconian Alb, wayside stones, heaps of stone which have been taken from and laid along the edges of fields; freshly ploughed fields; newly unearthed material which comes to light in the course of mechanical excavations; earthmoving during building or road construction; and smaller exposures and crevices in the earth along the edges of forests and paths.
J.P. Felix, who was born in Leipzig in 1859, was the first person to investigate the cellular structures of southern German Molasse woods by the use of thin sections. The items which he discussed in his 1882 dissertation "Über Studien versteinerter Hölzer" include specimens from Brazil, Cuba, Antigua, Asia, Japan, Java, and Australia, as well as woods found in Hungary, Prussia, Saxony, and Bavaria. He identified one "wood-stone" from Wagenhofen south of Neuburg on the Danube as *Sapotoxylon* sp., a member of the Sapindaceae family. The original specimens, which had been kept in Leipzig, was reexamined a century later and identified as Juglandacee wood type, i.e. a type of walnut.
K. Mägdefrau (1908-1999) encouraged his students to work on palaeobotanical themes. During subsequent decades, this encouragement led to intensive research into southern German fossil plants. With specialized assistance from W. Jung, students wrote numerous scientific articles about many newly found Molasse discovery sites and the plant remains that were found there (leaves, seeds, fruits, woods, pollen, and algae).
The oldest mention of petrified woods from strata of the southern German Molasse was given by the Benedictine Placidus Schärl in 1794. In a readable publication entitled

"Von Versteinerung des Holzes", many localities are mentioned where silicified woods can be found today. During the past few decades, "fossil hunters" have found in the Molasse, in addition to the remains of animals, thousands of silicified woods. In the area of the Upper Freshwater Molasse, there exist seven particularly rich sites, including so-called "drift-wood deposits." A few abundant sites of silicified woods are mentioned here: Collection L. Fruth – southern Franconian Alb; P. Holleis – southern Franconian Alb; L. Lang – the surroundings of Neuburg on the Danube; H. Jaksch – the region around Oberbachern and Radersdorf; H. Haberda – Danube gravel near Deggendorf; R. Baumgartner – Ortenburger Schotter in the Passau region. These finds are primarily kept at the Bavarian State Collection for Palaeontology and Geology in Munich, at the Augsburg Museum of Natural History, at the "Historical Association" in Neuburg on the Danube, as well as in various private collections.

Identified Woods

Thin sections, microscopes, comparative slides in a "xylothek", special literature, microscopic illustrations, and descriptions of regional woody floras, as well as knowledge of the woody anatomy, are all necessary to successfully identify Tertiary woods. Even the smallest "wood-stones," some of which may measure no more than a few centimeters, may be the first indication of the existence of a previously unknown genus. The largest silicified woods from the southern German Molasse rarely attain lengths of as much as 5 to 7 meters; the smallest pieces measure only a few centimeters. Microscopic examination of more than 600 specimens of Late Tertiary wood shows that the majority of these fossils are from deciduous trees. The most frequently identified taxa are: Lauraceae (*Laurus*, *Cinnamomum*), Leguminosae (*Acacia* structure, *Albizzia*, *Robinia*), Fagaceae (*Quercus*, *Castanea*, *Castanopsis*), Platanaceae, Juglandaceae, and Hamamelidaceae (*Liquidambar*). Dispersed, individual discoveries

9 *Cedreloxylon crystalliferum*
Cedrela "cigar-box wood", Meliaceae family, found at Seibersdorf am Inn, Lower Bavaria, Tangential section with whirled rays length of small ray 180 μm

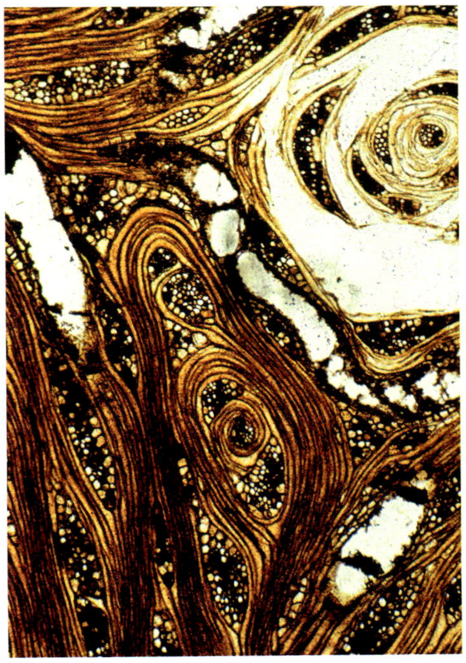

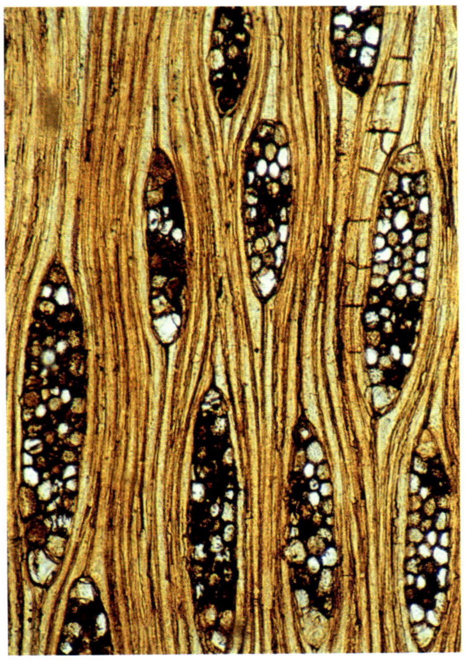

10 *Cedreloxylon crystalliferum*
Cedrela "cigar-box wood", Meliaceae family, found at Seibersdorf am Inn, Niederbayern. Lower Bavaria. Tangential section with spindle-shaped rays and rhomboidal individual crystals length of large ray 820 μm

within the southern German Molasse (not including the woods from the Ortenburger Schotters) belong to the families Anacardiaceae (*Pistacia*), Dipterocarpaceae (*Shorea*), Meliaceae (*Cedrela*), Moraceae (*Morus*), Sapotaceae (*Bumelia*), Rosaceae (*Crataegus*, *Prunus*), and Salicaceae (*Populus*). Silicified *Lithocarpus* (*Pansonia*) woods belong to the "evergreen species" of the *Quercus* section Cyclobalanus. They correspond to the wood of huge, recent, individual trees growing in SE Asia. More than 40, in part excellently preserved woods of the edible chestnut (*Castanea*) were found deposited in the coarse Tertiary gravel of the Neuburger Wald near Passau. More than 60 gymnospermous woods and some 20 *Palmoxylon* specimens round out the list of deciduous woods. A special feature are four discovery sites with the extinct Cretaceous fern *Tempskya*.

Woods From the "Ortenburg Gravel"

Limno-fluvial, terrestrial deposits characterized the landscape during deposition of the Upper Freshwater Molasse. The whitish-gray "Ortenburg Gravel" in the eastern part of the Bavarian Molasse Basin, which takes its name from the city of Ortenburg, was created from material deposited in deltas as the remaining ocean retreated from the Upper Marine Molasse. The neatly defined deposit is between 5 and 10 km wide and attains a maximum thickness of 50 m. Its age is believed to be 18.5 million years corresponding to the Lower Miocene. The quartz-rich pebbles and gravel carry alternating amounts of central Alpine sedimentary components. The "Ortenburg Gravel" has been known for decades among collectors because of its rich trove of fossils. Examination of approximately

11 *Cedreloxylon crystalliferum* Cedrela "cigar-box wood", Meliaceae family, found at Seibersdorf am Inn, Niederbayern. Lower Bavaria. Tangential section with rhomboidal individual crystals in ray cells; maximum diameter of the crystals 30 μm

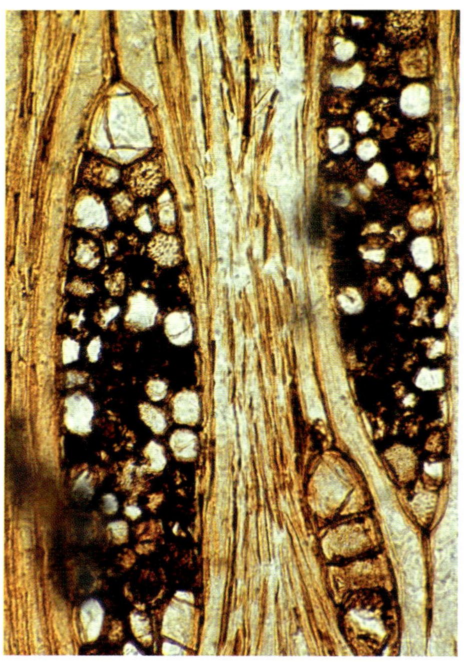

12 *Carapoxylon ortenburgense* Carapa, mahogany, Meliaceae family, found at Rauscheröd, Lower Bavaria. Radial section with fungal hyphae in water-conductive tissue; radial diameter of the conductive elements 190 μm

150 specimens has shown that these silicified woods are markedly different from those found at other comparable Tertiary sites. 56 specimens (i.e. roughly one-third of the total) belong to the widely distributed tropical Meliaceae family (*Carapa*, *Xylotropus*). These are mangroves that grow towards the land or trees that prefer the shores of coastal rivers. The Meliaceae are distributed throughout Asia, Australia, and East Africa. The living relatives of another 9 identified genera also inhabit a purely tropical area. They belong to the families Taxodiaceae, Bombacaceae, Ebenaceae, Ericaceae, Euphorbiaceae, Fagaceae, Flacourtiaceae, Lauraceae, Meliaceae, Myricaceae, Tiliaceae, and Arecaceae.
The woods were found in the large Rauscheröd Exposure, a gravel quarry located 5 km NW of Ortenburg.
The majority of woods are part of the Baumgartner Collection and are kept in the Bavarian State Collection for Palaeontology and Geology in Munich.

Late Tertiary Landscape

During the Late Tertiary, i.e. approximately 18 million years ago, the ocean finally departed from southern Germany, but the deposition of gravel continued here with the series of the Upper Freshwater Molasse. Material recovered from drilling cores has shown that these terrestrially deposited, Tertiary strata attain a thickness of up to 1,000 m at some sites. Long-term depressions of the southern German Molasse Basin with simultaneous sedimentation created these often very thick strata of sand and pebbles.
The Molasse, with its series of sediments accessible in many sand and pebble quarries, thus offers an especially valuable insight into the Central European habitat 18 to 11 million years ago. The extremely rich and exotic vertebrate fauna from Sandelzhausen near Mainburg and plant-rich brown-coal deposits from the Naab Molasse north of Regensburg are excellent examples of this. Anastomosing river systems on flood plains associated with lakes, swamps and repeatedly inundated riverine meadowlands formed the landscape during the era of the Upper Freshwater Molasse.
The early coverage of plant remains by riverine sand or lake mud prevented oxygen from entering and thus made it possible for the plants to become transformed into a fossilized state. Plant remains thus survived as fossils and as indicators of the climate. These can now be used as clues upon which to base a reconstruction of the ancient vegetation.

In the course of millions of years, the Tertiary Molasse forests with their exotic evergreen trees experienced, as a result of declining temperatures, a change which has also been documented at other sites elsewhere in the world. To an increasingly great extent, woody plants which remained green during the summer and which lost their leaves during the winter wandered into this habitat.
The forests gradually began to resemble those which are currently found in the temperate latitudes of the Northern Hemisphere.

Finally, during the Pleistocene Ice Age, the species-rich Tertiary forests were decimated and their remnants receded from large areas of Europe. The rich discoveries of fossilized woods in the Molasse strata of the Alpine foothills allow these forests to rise again before our mind's eye.

References

Rößler: Study Methods for Determining the Structure of Plant Organs

Bachofen von Echt, C. 1867. Versuch chemischer Analysen einiger Holzversteinerungen.- In: Corda, A.J.: Flora Protogaea - Beiträge zur Flora der Vorwelt, p. 11-14; Berlin (S. Calvary & Co.).

Dietrich, D.; Frosch, G.; Rößler, R. & Marx, G. 2000. Analytical X-ray microscopy on *Psaronius* sp. - a contribution to the permineralization process.- Mikrochimica Acta, 133: 279-283.

Dietrich, D.; Witke, K.; Rößler, R. & Marx, G. 2001. Raman spectroscopy on Psaronius sp.: a contribution to the understanding of the permineralization process.– Applied Surface Science, 179: 230-233.

Götze, J. & Rößler, R. 2000. Kathodolumineszenz-Untersuchungen an Kieselhölzern aus dem Perm von Chemnitz.– Veröff. Mus. Naturk. Chemnitz, 23: 35-50.

Jones, T.P. & Rowe, N.P. (Eds.) 1999. Fossil Plants and Spores: modern techniques. Geological Society, London, 1-396 pp.

Rößler, R. 2001. Die Versteinerung von Hölzern – ein ewiges Rätsel?– In: Rößler, R. (Ed.): Der versteinerte Wald von Chemnitz. Katalog zur Ausstellung Sterzeleanum, p. 192-205; Chemnitz.

Selmeier, A. 1994. Nadelholz oder Laubholz, Palme oder Farn? Wie man versteinertes Holz bestimmt.– extraLapis, 7: 32-47.

Kerp: The Rhynie Chert - The Oldest and Most Completely Preserved Terrestrial Ecosystem

Edwards, D. & Selden, P.A. 1993. The development of Early terrestrial ecosystems.- Bot. J. Scotland, 46: 337-366.

Edwards, D.; Kerp, H. & Hass, H. 1998. Stomata in early land plants: an anatomical and ecophysiological approach.- J. Exp. Bot., 49: 255-278.

Hass, H. & Remy, W. 1992. Devonian fungi. Interactions with the green algae Palaeonitella.- Mycologia, 86: 901-910.

Hass, H.; Taylor, T.N. & Remy, W. 1994. Fungi from the Lower Devonian Rhynie chert: Mycoparasitism.- Amer. J. Bot., 81: 29-37.

Hirst, S. 1923. On some arachnid remains from the Old Red Sandstone (Rhynie Chert Bed, Aberdeenshire).- Ann. Mag. nat. Hist., 9th Ser., 70: 455-474.

Kidston, R. & Lang, W.H. 1917. On Old Red Sandstone plants showing structure, from the Rhynie chert bed, Aberdeenshire. Part I. *Rhynia gwynne-vaughani* Kidston & Lang.- Trans. Roy. Soc. Edinburgh, 51(24): 761- 784.

Kidston, R. & Lang, W.H. 1920. On Old Red Sandstone plants showing structure, from the Rhynie Chert bed, Aberdeenshire. Part II. Additional notes on *Rhynia gwynne-vaughani*, Kidston and Lang; with descriptions of *Rhynia major*, n. sp., and *Hornia lignieri*, n.g., n.sp.- Trans. Roy. Soc. Edinburgh, 52(24): 603-627.

Kidston, R. & Lang, W.H. 1920. On Old Red Sandstone plants showing structure, from the Rhynie Chert bed, Aberdeenshire. Part III. *Asteroxlon Mackiei*, Kidston and Lang.- Trans. Roy. Soc. Edinburgh, 52(26): 643-680.

Kidston, R. & Lang, W.H. 1921. On Old Red Sandstone plants showing structure, from the Rhynie Chert bed, Aberdeenshire. Part IV. Restorations of the vascular cryptogams, and discussion of their on the general morphology of the pteridophyta and the origin of the organisation of land-plants.- Trans. Roy. Soc. Edinburgh, 52(32): 831-854.

Kidston, R. & Lang, W.H. 1921. On Old Red Sandstone plants showing structure, from the Rhynie Cert bed, Aberdeenshire. Part V. The Thallophyta occuring in the peat-bed; the succession of the plants throughout a vertical section of the bed, and the conditions of accumulation and preservation of the deposit.- Trans. Roy. Soc Edinburgh, 52(33): 855-902.

Mackie, W. 1916. The Rock series of Craigbeg and Ord Hill, Rhynie, Aberdeenshire.- Trans. Edinburgh Geol. Soc., 10: 205-236.

Remy, W.; Remy, D. & Hass, H. 1997. Organisation, Wuchsformen und Lebenstrategien früher Landpfanzen des Unterdevons.- Bot. Jahrb. Syst., 19: 509-562.

Shear, W.A. 1991. The early development of terrestrial ecosystems.- Nature, 351: 283-289.

Taylor, T.N.; Hass, H. & Kerp, H. 1999. The oldest fossil ascomycetes.- Nature, 399: 648.

Taylor, T.N.; Hass, H., Remy, W. & Kerp, H. 1995. The oldest fossil lichen.- Nature, 378: 244.

Taylor, T.N.; Remy, W. & Hass, H. 1994. Allomyces in the Devonian.- Nature, 367: 601.

Trewin, N.H. & Rice, C.M. 1992. Stratigraphy and sedimentology of the Devonian Rhynie chert locality.- Scott. J. Geol., 28: 37-47.

Galtier: Exceptionally Preserved Lower Carboniferous Plants

Renault, B. 1893-1896. Bassin Houiller et permien d'Autun et d'Epinac. 4. Flore fossile. 2. Etude des Gites Minéraux de la France, 578 pp.

Rex. G. 1986. The preservation and palaeoecology of the Lower Carboniferous silicified plant deposits at Esnost, near Autun, France.- Geobios, 19: 773-800.

Rex, G. & Scott, A.C. 1987. The sedimentology, palaeoecology and preservation of the Lower Carboniferous plant deposits at Pettycur, Fife, Scotland.- Geol. Mag. 124: 43-66.

Rolfe, W.D.I.; Clarkson, E.N.K. & Panchen, A.L. 1994. Volcanism and early terrestrial biotas.- Trans. Roy. Soc. Edinburgh, Earth Sci., 84: 175-464.

Williamson, W.C. 1871-1895. On the organization of the fossil plants of the Coal-Measures. Parts 1-19.- Phil. Trans. Royal Soc. London. 161-194.

Galtier: *Pitus*, a Giant Tree of the Early Carboniferous Time

Long, A.G.1979. Observations on the Lower Carboniferous genus *Pitus* Whitham.- Trans. Roy. Soc. Edinburgh, Earth Sci., 70: 111-127.

Witham, H.T.M. 1830. Vegetation of the first period of an ancient world. Phil. Mag. 7:23-31.

Witham, H.T.M. 1833. The internal structure of fossil vegetables found in the Carboniferous and oolitic deposits of Great Britain. Edinburgh: A&C Black, 84 pp.

Rothwell: Coal Balls - Remarkable Evidence of Palaeozoic Plants and the Communities in Which They Grew

Bateman, R.M.; DiMichele, W.A. & Willard, D.A. 1992. Experimental cladistic analysis of anatomically preserved arborescent lycopsids from the Carboniferous of Euramerica: an essay on paleobotanical phylogenetics.- Annals of the Missouri Botanical Garden, 79: 500-559.

Beeler, H.E. 1983. Anatomy and frond architecture of *Neuropteris ovata* and *N. scheuchzeri* from the Upper Pennsylvanium of the Appalachian Basin.- Canadian Journal of Botany, 61: 2352-2368.

Bract, S.D. 1970. On a new structurally preserved arborescent lycopsid fructification from the Lower Pennsylvanium of North America.- American Journal of Botany. 57: 317-330.

Binney, E.W. 1868. On fossil wood in calcareous nodules found in the upper foot coal near Oldeham.- Proceedings of the Philosophical and Literary Society of Manchester, 5: 113.

Brack-Hanes, S.D. & Vaughn, J.C. 1978. Evidence of Paleozoic chromosomes from lycopod microgametophytes.- Science, 200: 1383-1385.

Delevoryas, T. 1964. Ontogenetic studies of fossil plants.- Phytomorphology, 14: 299-314.

DiMichele, W.A. 1980. *Paralycopodites* Morey and Morey, from the Carboniferous of Euramerica - a reassessment of the generic affinities and evolution of *„Lepidodendron" brevifolium* Williamson.- American Journal of Botany, 67: 1466-1476.

DiMichele, W.A. & Hook, R.W. (Rapporteurs). 1992. Paleozoic terrestrial ecosystsems. S. 206-325. In: Behrensmeyer, A.K.; Damuth, D.; DiMichele, W.A.; Potts, R.; Sues, H.D. & Wing, S.L. (Eds.): Terrestrial ecosystems through Time. University of Chicago Press, Chicago.

DiMichele, W.A.; Pfefferkorn, H.W. & Phillips, T.L. 1996. Persistence of Late Carboniferous tropical vegetation during glacially driven climatic sea-level fluctuations.- Palaeogeography, Palaeoclimatology, Palaeoecology, 125: 105-128.

DiMichele, W.A. & Phillips, T.L. 1996. Climate change, plant extinctions and vegetational recovery during the Middle-Late Pennsylvanium Transition: the case of tropical peat-forming environments in North America. pp. 201-221. In: Hart, M.B. (Ed.): Biotic recovery from mass extinction events.- Geological Society Special Publication, 102.

Eggert, D.A. 1961. The ontogeny of the Carboniferous arborescent Lycopsida.- Palaeontographica, 108B: 43-92.

Gastaldo, R.A.; DiMichele, W.A. & Pfefferkorn, H.W. Out of the icehouse into the greenhouse: a Late Paleozoic analog for modern global vegetational change.- GSA Today, 6: 1-7.

Good, C.W. 1971. The ontogeny of Carboniferous articulates: calamite leaves and twigs.- Palaeontographica, 133B: 137-158.

Good, C.W. & Taylor, T.N. 1974. The establishment of *Elaterites triferens* spores in *Calamocarpon insignis* microsporangia.- Transactions of the American Microscope Society, 93: 148-151.

Joy, K.W.; Willis, A.J. & Lacey, W.S. 1956. A rapid cellulose acetate peel technique in paleobotany.- Annals of Botany, 20: 635-637.

Leclercq, S. 1925. Les Coal Balls de la couche Bouxharmont des Charbonnages de Wérister.- Mémoires de la Société Géologique de Belgique, Liége, 79 pp.

Li, H.; Tian, B.; Taylor, E.L. & Taylor, T.N. 1994. Foliar anatomy of *Gigantoclea guizhouensis* (Gigantopteridales) from the Upper Permian of Guizhou Province, China.- American Journal of Botany, 81: 678-689.

Millay, M.A. & Eggert, D.A. 1974. Microgametophyte development in the Paleozoic seed fern family Callistophytaceae.- American Journal of Botany, 61: 1067-1075.

Phillips, T.L. 1980. Stratigraphic and geographic occurrences of permineralized coal-swamp plants - Upper Carboniferous of North America and Europe. pp. 25-92. In: Dilcher, D.L. & Taylor, T.N. (Eds.): Biostratigraphy of fossil plants. Dowden, Hutchinson and Ross, Inc., Stroudsburg, Pennsylvania.

Phillips, T.L. & DiMichele, W.A. 1981. Paleoecology of Middle Pennsylvanium age coal swamps in Southern Illinois; Herrin Coal member at Sahara Mine No. 6. S. 231-285. In: Niklas, K.J. (Ed.): Paleobotany, Paleoecology and Evolution. Praeger, Vol I., New York.

Phillips, T.L.; Peppers, R.A. & DiMichele, W.A. 1985. Stratigraphic and interregional changes in Pennsylvanium coal-swamp vegetation: environmental infrences.- International Journal of Coal Geology, 5: 43-109.

Pigg, K.B.; Sophy, J. & Wehr, W.C. 1996. Petrified Miocene flora from Yakima Canyon, Washington, USA.- Abstracts of the International Organisation of Palaeobotany Conference V, Santa Barbara, p. 80.

Pigg, K.B. & Stockey, R.A. 1996. The significance of the Princeton Chert permineralized flora to the Middle Eocene upland biota of the Okanogan Highlands.- Washington Geology, 24: 32-36.

Pryor, J.S. 1996. The Upper Pennsylvanium Duquesne Coal of Ohio (USA): Evidence for a dynamic peat-accumulating swamp community.- International Journal of Coal Geology, 29: 119-146.

Remy, W. & Hass, H. 1991. Gametophyten und Sporophyten im Unterdevon - Fakten und Spekulationen.- Argumenta Palaeobotanica, 8: 193-223.

Rößler, R. 2000. The late Paleozoic tree fern *Psaronius* - an ecosystem unto itself.- Review of Palaeobotany and Palynology, 108: 55-74.

Rothwell, G.W. 1976. Petrified Pennsylvanium age plants of eastern Ohio.- Ohio Journal of Science, 76: 128-132.

Rothwell, G.W. 1981. The Callistophytales (Pteridospermopsida); reproductively sophisticated Paleozoic gymnosperms.-

Review of Palaeobotany and Palynology, 32: 103-121.
Rothwell, G.W. 1996. Pteridophytic evolution: an often underappreciated phytological success story.- Review of Palaeobotany and Palynology, 90: 209-222.
Rothwell, G.W. 1999. Fossils and ferns in the resolution of land plant phylogeny.- The Botanical Review, 65: 188-218.
Rothwell, G.W. & Erwin, D.M. 1985. The rhizomorph apex of *Paurodendron*; implications for homologies among the rooting organs of *Lycopsida*.- American Journal of Botany, 72: 86-98.
Rothwell, G.W. & Good, C.W. 2000. Reconstructing the Pennsylvanium-age filicalean fern *Botryopteris tridentata* (Felix) Scott.- International Journal of Plant Sciences, 161: 495-507.
Rothwell, G.W. & Warner, S. 1984. *Cordaixylon dumusum* n. sp. (Cordaitales). I. Vegetative structures.- Botanical Gazette, 145: 275-291.
Sahni, B. 1931. On certain fossil epiphytic ferns found on stems of the Palaeozoic tree-fern *Psaronius*. p. 270.- In: Proceedings of the 18th Indian Science Congress, Nagpur, India.
Sahni, B. 1948. The Pentoxyleae: A new group of Jurassic gymnosperms from the Rajmahal Hills of India.- The Botanical Gazette, 110: 47-80.
Stewart, W.N. 1947. A comparative study of stigmarian appendages and *Isoetes* roots.- American Journal of Botany, 34: 315-324.
Stewart, W.N. & Rothwell, G.W. 1993. Paleobotany and the evolution of plants. 2nd Ed., Cambridge University Press, Cambridge, 521 pp.
Stubblefield, S.B. & Rothwell, G.W. 1981. Embryogeny and reproductive biology of *Bothrodendrostrobuts mundus* (Lycopsida).- American Journal of Botany, 68: 625-634.
Taylor, T.N. & Eggert, D.A. 1969. Preliminary study of spores from Carboniferous fructifications utilizing the scanning electron microscope. p. 97-107. In the Proceedings of the Engis Steroscan Colloquium, 1969. Engis Equipment Company, Austin.
Taylor, T.N. & Taylor, E.L. 1989. Antarctic Paleobiology. Springer-Verlag, New York, 261 pp.
Taylor, T.N. & Taylor, E.L. 1993. The biology and evolution of fossil plants. Prentice Hall, Englewood Cliffs, New Jersey, 982 p.
Tian, B.; Wang, S.J.; Guo, Y.; Chen, G. and Zhao, H. 1996. Flora of Palaeozoic Coal balls of China.- Paleobotanist, 45: 247-254.
Trivett, M.L. 1992. Growth Architecture, structure, and relationships of *Cordaixylon iowensis* nov. comb. (Cordaitales).- International Journal of Plant Sciences, 153: 273-287.
Trivett, M.L. & Rothwell, G.W. 1991. Diversity among Palaeozoic Cordaitales.- Neues Jahrb. Geol. Palaeont. Abh. B, 183: 289-305.

Rößler: Rößler: The Petrified Trees From Kyffhäuser Mountain, Germany

Gaitzsch, B.G. 2001. Das Oberkarbonprofil der Saale-Senke am Kyffhäuser - Eine Diskussion.- Beiträge zur Geologie von Thüringen, N.F. 8: 43-66; Jena.
Göppert, H.R. 1880. Ueber die versteinerten Hölzer des Kyffhäuser.- Neues Jahrbuch für Mineralogie, Geologie und Palaeontologie, 2: 89-92.
Mägdefrau, K. 1958. Die Kieselhölzer im obersten Oberkarbon des Kyffhäuser-Gebirges.- Berichte der Deutschen Botanischen Gesellschaft, 71: 133-142.
Mickle, J.E. & Barthel, M. 1992. *Psaronius*-Stämme im Oberkarbon des Kyffhäuser?- Abhandlungen und Berichte des Museums der Natur Gotha, 17: 11-14.
Weiß, H.-J. 1998. Beobachtungen an Kieselhölzern des Kyffhäuser-Gebirges.- Veröffentlichungen des Museums für Naturkunde Chemnitz, 21: 37-48.
Weiß, H.-J. & Noll, R. 1999. „Holzkristalle"- Veröffentlichungen des Museums für Naturkunde Chemnitz, 22: 57-64.

Rößler: Psaronius Forests at the Feet of Volcanoes

Barthel, M. & Rößler, R. 1997. Tiefschwarze Kieselstämme aus Manebach.- Abh. Naturhist. Mus. Schloß Bertholdsburg Schleusingen, 12: 53-61.
Galtier, J. & Phillips, T.L. 1979. Swamp vegetation from Grand'Croix (Stephanian) and Autun (Autunian), France, and Comparisons with Coal-Ball Peats of the Illinois Basin. – 9eme Congr. Int. De Strat. Géol. Carbonif., Washington and Champaign-Urbana, Compte Rendu, 5: 13-24.
Göppert, H.R. 1859. Ueber die versteinerten Wälder im nördlichen Böhmen und in Schlesien.- Verhandlungen der schlesischen Gesellschaft Breslau, 1858/59: 1-11.
Göppert, H.R. 1864-1865. Die fossile Flora der permischen Formation. In: Meyer, H. von (Ed.): Palaeontographica-Beiträge zur Naturgeschichte der Vorwelt, Band VII: 316 p.; Cassel (Th. Fischer Verlag).
Renault, M.B. 1893/96. Bassin houiller et permien d'Autun et d'Epinac. Fasc. IV, Flore Fossile (2e partie).– Ét. Gites minéraux France. Atlas 1893, Text 1896: 578 p.; Paris.
Rößler, R. 1996. Explosiver Vulkanismus im Oberrotliegenden - Die Überlieferungschance des „Versteinerten Waldes" von Chemnitz. - Natur und Museum, 126: 431-439.
Sterzel, J.T. 1875. Die fossilen Pflanzen des Rothliegenden von Chemnitz in der Geschichte der Palaeontologie.- Ber. Naturwiss. Ges. Chemnitz, 5: 71-243.
Sterzel, J.T. 1918. Die organischen Reste des Kulms und des Rotliegenden der Gegend von Chemnitz.– Abh. Königl. Sächs. Ges. Wiss., Math.-phys. Kl., 35, 5: 205-315; Leipzig (B.G. Teubner).

Barthel: The Maggot Stones From Windberg Ridge, Germany

Barthel, M.; Reichel, W. & Weiss, H.-J. 1995. „Madensteine" in Sachsen. Neue Funde von *Scolecopteris elegans* in der Typuslokalität.- Abhandlungen des Staatlichen Museums für Mineralogie und Geologie zu Dresden, 41: 117-135.

Barthel, M. & Weiss, H.-J. 1997. Xeromorphe Baumfarne im Rotliegend Sachsens.- Veröffentlichungen des Museums für Naturkunde Chemnitz 20: 45-56.

Millay, M.A. 1979. Studies of Palaeozoic Marattialeans: A monograph of the American species of *Scolecopteris*.- Palaeontographica, B 169: 11-69.

Millay, M.A. 1997. A review of permineralized Euramerican Carboniferous tree Ferns.- Review of Palaeobotany and Palynology, 95: 1191-1209.

Reichel, W. & Schauer, M. i. Druck. Das Döhlener Becken. Bergbau-Monographie.- In: Bergbau in Sachsen Bd. 9, Sächsisches Landesamt für Umwelt und Geologie und Bergamt Freiberg.

Solms-Laubach, H. von 1883. Zur Geschichte der *Scolecopteris elegans* Zenker.- Nachrichten der Königlichen Gesellschaft der Wissenschaften und der Georg-August-Universität zu Göttingen, 2: 26-38.

Sterzel, J.T. 1893. Die Flora des Rothliegenden im Plauenschen Grunde bei Dresden.- Abhandlungen der Königlich-Sächsischen Gesellschaft der Wissenschaften, 32: 1-172.

Strasburger, E. 1874. Über *Scolecopteris elegans* Zenk., einen fossilen Farn aus der Gruppe der Marattiaceen.- Jenaische Zeitschrift für Naturwissenschaften, 8: 81-95.

Dernbach, Noll & Rößler: News From Araguaina, Brazil

Rößler, R. & Galtier, J. 2002a. First *Grammatopteris* tree ferns from the Southern Hemisphere - new insights in the evolution of the Osmundaceae from the Permian of Brazil. - Rev. Palaeobot. Palynol., Amsterdam (Elsevier), in press.

Rößler, R. & Galtier, J. 2002b. *Dernbachia brasiliensis* gen. nov. et sp. nov. - a new small tree fern from the Permian of NE-Brazil. - Rev. Palaeobot. Palynol., Amsterdam (Elsevier), in press.

Rößler, R. & Noll, R. 2002. Der permische versteinerte Wald von Araguaina/Brasilien - Geologie, Taphonomie und Fossilführung.- Veröff. Museum für Naturkunde Chemnitz, 25, in press.

Noll & Wilde: Conifers from the "Uplands" - Petrified Wood from Central Germany

Barthel, M., Reichel, W. & Weiss, H.-J. 1995. „Madensteine" in Sachsen, neue Funde von *Solecopteris elegans* Zenker in der Typus-Lokalität.- Abh. Staatl. Mus. Mineral. Geol. Dresden, 41: 117-135; Dresden.

Barthel, M., Bettag, E. & Noll, R. 1998. *Dicranophyllum hallei* Remy & Remy im oberen Rotliegend.- Veröff. Mus. Naturkunde Chemnitz, 21: 5-20; Chemnitz.

Frentzen, K. 1931. Beiträge zum Oberrheinischen Fossilkatalog Nr. 6. Studien über die fossilen Hölzer der Sammelgattung Dadoxylon Endl. II. Teil. Untersuchung von Dadoxyla aus dem Karbon und dem Perm Europas mit besonderer Berücksichtigung der Funde aus dem Oberrheingebiete.- Abh. Heidelberger Akad. Wiss., Math.- naturwiss. Klasse, 19: 1-46; Berlin, Leipzig.

Göppert, H.R. 1864/65. Die fossile Flora der Permischen Formation. - Palaeontographica, 12(1864-1865): 1-316; Cassel.

Goth, K. & Wilde, V. 1989. Der versteinerte Wald von Kilianstädten.- Natur u. Museum, 119(7): 211-219; Frankfurt am Main.

Gothan, W. 1905. Zur Anatomie lebender und fossiler Gymnospermen-Hölzer.- Abh. der Königlich Preußischen Geol. L.-Anst., N. F., 44: 1-108; Berlin.

Haneke, J. 1998. In die Erde geschaut: Ergebnisse von sechs geologischen Forschungsbohrungen im Rotliegend des Donnersbergkreises. - Schr. Kreisvolkshochschule Donnersbergkreis, 2: 10-24; Kirchheimbolanden.

Henningsen, D. & Katzung, G. 1992. Einführung in die Geologie Deutschlands. 4. neu bearbeitete u. erw. Aufl., 228 p.; Stuttgart (Deutscher Taschenbuch Verlag u. Ferdinand Enke Verlag).

Keller, T. & Schmidt, K. 1997. Der versteinerte Wald von Kilianstädten, Bericht über eine Notbergung.- Denkmalpflege in Hessen, 1997(1): 21-25; Wiesbaden.

Noll, R., Wilde, V. & Goth, K. 1999. Kieselhölzer mit *Tylodendron*-Mark aus dem Pfälzischen Rotliegenden.- Natur u. Museum, 129(2): 45-57; Frankfurt am Main.

Potonié, H. 1888. Die fossile Pflanzen-Gattung *Tylodendron*.- Jahrb. der Königlich Preußischen Geol. Landesanst. u. Bergakad., 1887: 311-331; Berlin.

Rößler, R. 2001. Der versteinerte Wald von Chemnitz. 1-251 pp.; Chemnitz (Museum für Naturkunde).

Schaarschmidt, F. 1973. Die Kieselhölzer von Bad Vilbel - Zeugen eines Floren- und Klimawandels im ausgehenden Erdaltertum.- Bad Vilbeler Heimatblätter, 8: 3-14; Bad Vilbel.

Schuster, J. 1908. Kieselhölzer der Steinkohlenformation und des Rotliegenden aus der bayerischen Rheinpfalz.- Geognostische Jahresh., 20 (1907): 1-16; München.

Weiss, C.E. 1869-1872. Fossile Flora der jüngsten Steinkohlenformation und des Rothliegenden im Saar-Rhein-Gebiete.- 1-250; Bonn (A. Henry).- [Die *Tylodendron* betreffende Lieferung erschien 1871]

Rößler: Between Precious Inheritance and Immediate Experience – Palaeobotanical Research ...

Barthel, M. 1976. Die Rotliegendflora Sachsens.- Abh. Staatl. Mus. Min. Geol., 24: 1-190; Dresden.

Bertrand, P. 1911. Structure des Stipes d'*Asterochlaena laxa* Stenzel.- Mém. de la Soc. Géol. Du Nord, Tome 7, 1: 72 pp.

Bertrand, C. & Bertrand, P. 1911. Le *Tubicaulis berthieri* nov. sp.- Bull. Soc. Hist. Nat., 24: 43-92.

Corda, A.J. 1845. Beiträge zur Flora der Vorwelt. 128 p.; Prag (Calve'sche Buchhandlung).

Cotta, B. v. 1832. Die Dendrolithen in Bezug auf ihren inneren Bau. 89 p.; Dresden und Leipzig (Arnoldische Buchhandlung).

Frenzel, D. 1749. Kurtzer Bericht von denen versteinerten Höltzern und anderen natürlichen Seltenheiten um Chemnitz.- In: Grundig, C.G. (Ed.): Neue Versuche nützlicher Sammlungen zu der Natur- und Kunstgeschichte sonderlich von Ober-Sachsen. Bd. 1: 505-510; Schneeberg.
Göppert, H.R. & Stenzel, C.G. 1881. Die Medulloseae. Eine neue Gruppe der fossilen Cycadeen. p. 1-17; Kassel (Theodor Fischer).
Kerp, H. & Krings, M. 1998. Climbing and scrambling growth habits: common life strategies among Late Carboniferous seed ferns.- C. R. Acad. Sci. Paris, Earth & Planet. Sci., 326: 583-588.
Rößler, R. 2000. The late Paleozoic tree fern Psaronius - an ecosystem unto itself. - Rev. Palaeobot. Palynol., 108: 55-74.
Rößler, R. (Ed.) 2001. Der versteinerte Wald von Chemnitz. The Petrified Forest of Chemnitz. Katalog zur Ausstellung Sterzeleanum, 253 pp.; Chemnitz (Museum für Naturkunde).
Sahni, B. 1932. On a palaeozoic tree-fern, *Grammatopteris baldaufi* (Beck) Hirmer, a link between the Zygopterideae and Osmundaceae.- Annals of Botany, XLVI, CLXXXIV: 863-879.
Sprengel, A. 1828. Commentatio de Psarolithis Ligni Fossilis Genere. 38 p.; Halle.
Sterzel, J.T. 1887. Über den großen *Psaronius* in der Naturwissenschaftlichen Sammlung der Stadt Chemnitz.– Ber. Naturwiss. Ges. Chemnitz, 10: 144-156.
Urban, G. 1970. David Frenzel - Sächsischer Edelgestein-Inspektor zu Chemnitz.- Veröff. Mus. Naturk. Karl-Marx-Stadt, 5: 3-7.
Weber, O. & Sterzel, J.T. 1896. Beiträge zur Kenntnis der Medulloseae.- Ber. Naturwiss. Ges. Chemnitz, 13: 44-143.
Zeiller, R. 1888. Études sur le terrain houiller de Comentry. Flore Fossile. Prem. Part, Saint-Étienne, 366 p.

Tidwell: Osmundaceae, Guaireaceae, Tempskyaceae, Cycadophyten, Hermanophyton, Extant Tree Ferns ...

Andrews, H.N., Jr. 1950. A fossil osmundaceous tree-fern from Brazil.- Bull. Torrey Bot. Club, 77(1): 29-34.
Andrews, H.N., & Kern, E.M. 1947. The Idaho *Tempskya*'s and associated fossil plants.- Missouri Bot. Gard. Ann., 34(2): 119-183.
Archangelsky, S. & D.W. Brett 1963. Studies on Triassic fossil plants from Argentina. II. *Michelilloa waltonii* nov. gen. et spec. from the Ischigualasto Formation.- Ann. Bot., 27: 147-154.
Archangelsky, S. & D.W. Brett 1961. Triassic fossil plants from Argentina. I. *Rhexoxylon* from the Ischigualasto Formation.- Phil. Trans. Roy. Soc. London, B 224: 1-19.
Arnold, C.A. 1945. Silicified plant remains from the Mesozoic and Tertiary of western North America. I. Ferns.- Michigan Acad. Sci. Arts Lett. Pap., (1944) 30: 3-34.
Arnold, C.A. 1947. An Introduction to Paleobotany.- McGraw-Hill Book Co., New York. 433 pp.
Arnold, C.A. 1952. Fossil Osmundaceae from the Eocene of Oregon.- Palaeontographica, B 92: 63-78.
Arnold, C.A. 1958. A new *Tempskya*.- Contrib. Mus. Paleont., Univ. Michigan, 14: 133-142.
Arnold, C.A. 1962. A *Rhexoxylon*-like stem from the Morrison Formation of Utah.- Amer. J. Bot., 49: 833-836.
Arnold, C.A & Daugherty, L.H. 1963. The fern genus Acrostichum in the Eocene Clarno Formation of Oregon.- Contrib. Mus. Paleont., Univ. Michigan, 18: 205-227.
Arnold, C.A. & Daugherty, L.H. 1964. A fossil dennstaedtioid fern from the Eocene Clarno Formation of Oregon.- Contrib. Mus. Paleont., Univ. Michigan, 19: 65-88.
Ash, S.R. 1985. A short thick cycad stem from the Upper Triassic of Petrified Forest National Park, Arizona and vicinity.- Mus. Northern Arizona Bull., 54:17-32.
Ash, S.R. 1994. *Donwellicaulis chlouberii* gen. et. sp. nov. (Guaireaceae, Osmundales) one of the oldest plant megafossils known in North America.- Palaeontographica, B 234: 1-17.
Ash, S.R. 2000. Evidence of oribatid mite herbivory in the stem of a late Triassic tree fern from Arizona.- J. Paleont., 74(6): 1065-1071.
Ash, S.R. and C.B. Read. 1976. North American species of *Tempskya* and their stratigraphic significance.- U.S. Geol. Surv. Prof. Pap., 874: 42 p.
Bancroft, H. 1932. A fossil cyatheoid stem from Mount Elgon, East Africa.- New Phytol., 31: 241-253.
Barale, G. and I.L. Viera. 1989. *Tempskya riojana* nov. sp., Fougère Aborescente du Crètacè Infèrieur du Nord-quest de L'espange.- Palaeontographica, B 212: 103-122.
Barrington, D.S. 1983. *Cibotium oregonense*: an Eocene tree-fern stem and petioles with internal structure.- Amer. J. Bot., 70: 1118-1124.
Boodle, L.A. 1895. Spores in a specimen of *Tempskya* (Endogenites).- Ann. Bot., 9: 137-141.
Brown, R.W. 1936. Field identification of the fossil ferns called *Tempskya*.- Jour. Washington Acad. Sci., 26(2): 45-52.
Chandler, M.E.J. 1968. A new *Tempskya* from Kent.- London, British Mus. (Nat. History) Bull., Geology, 15(4): 169-179.
Corda, A.J. 1845. Flora Protogaea-Beiträge zur Flora der Vorwelt. Calvary, Berlin, 128 pp.
Crepet, W.L. 1974. Investigations of North American cycadeoids: The reproductive biology of Cycadeoidea.- Palaeontographica, B 148:144-169.
Daugherty, L.H. 1941. The Upper Triassic flora of Arizona.- Carnegie Inst. Washington Publ., 256: 1-108.
Daugherty, L.H. 1960. *Itopsidema*, a new genus of the Osmundaceae from the Triassic of Arizona.- Amer. J. Bot., 47: 771-777.
Erasmus, T. 1978. The anatomy and evolution of *Osmundacaulis* Miller emend. with notes on the geometry of the xylem framework of the osmundaceous stele.- Univ. Pretoria Unpubl. Ph.D. Dissertation, 155 pp.
Gould, R.E. 1970. *Palaeosmunda*, a new genus of siphonostelic osmundaceous trunks from the Upper Permian of Queensland.- Palaeontology, 13: 10-28.
Gould, R.E. 1971. *Lyssoxylon grigsbyi*, a cycad trunk from the Upper Triassic of Arizona and New Mexico.- Amer. J. Bot., 58:

239-248.
Gould, R.E. 1972. *Cibotum tasmanense* sp. nov., a fossil tree-fern from the Tertiary of Tasmania.- Austr. J. Bot., 20: 119-126.
Gould, R.E. 1973. A new species of *Osmundacaulis* from the Jurassic of Queensland.- Proc. Linnean Soc. New South Wales, 98(2): 86-94.
Gupta, K.M. 1971. Investigations on the Jurassic flora of the Rajmahal Hills, India. 9. On the structure and affinities of *Sewardioxylon sahnii* gen. et sp. nov., a petrified cycadean wood from the Rajmahal Hills, India.- Palaeontographica, B 131:160-166.
Herbst, R. 1981. *Guairea milleri* nov. gen. et sp. y Guaireaceae, nueva familia de las Osmundales (sensu lato) del Permico Superior de Paraguay.- Ameghiniana, 1-2: 35-50.
Herbst, R. 1986. Studies on Psaroniaceae. I. The Family Psaroniaceae (Marattiales) and a redescription of *Tietea singularis* Solms-Laubach, from the Permian and Brazil.- Actas Congr. Argentino Paleont. Bioestrat., 1(1986):163-171.
Herbst, R. 1999. Studies on Psaroniaceae. IV. Two species of *Psaronius* from Araguaina, state of Tocantis, Brazil.- Facena, 15: 9-18
Hewiston, W. 1962. Comparative morphology of the Osmundaceae.- Missouri Bot. Gard. Ann., 48: 57-93.
Jain, K.P. 1964. *Fascisvarioxylon methae* gen. et sp. nov., a new petrified cycadean wood from the Rajmahal Hills, Bihar, India.- Palaeobotanist, 11: 138-143.
Jung, W.W. 1983. *Tempskya*-Funde aus jungquartären Schottern Süddeutschlands.- Mitt. Bayer. Staatsslg. Paläont. Hist. Geol., 23:119-124.
Kidston, R. & Gwynne-Vaughan, D.T. 1908. On the fossil Osmundaceae. Part II.- Trans. Roy. Soc. Edinburgh, 46: 213-232.
Kidston, R. & Gwynne-Vaughan, D.T. 1911. On a new species of *Tempskya* from Russia.- Russ. k. min. Gesell. Verh., 48: 1-20.
Kidston, R. & Gwynne-Vaughan, D.T. 1914. On the fossil Osmundaceae. Part V.- Trans. Roy. Soc. Edinburgh, 50: 469-480.
Lantz, T.C., G.W. Rothwell, & R.A. Stockey. 1999. *Conantiopteris schuchmanii*, gen. et sp. nov., and the role of fossils in resolving the phylogeny of Cyatheaceae.- J. Plant Res., 112: 361-381.
Menendez, C.A. 1961. Estipite petrificado de una neuva Cyatheaceae del Terciario de Neuquén.- Bol. Soc. Argent. Bot., 9: 331-358.
Miller, C.N., JR. 1967. Evolution of the fern genus *Osmunda*.- Contrib. Mus. Paleont., Univ. Michigan, 21: 139-203.
Miller, C.N., JR. 1971. Evolution of the fern family Osmundaceae based on anatomical studies.- Contrib. Mus. Paleont., Univ. Michigan, 28: 105-169.
Miller, C.N., JR. 1982. *Osmunda wehrii*, a new species based on petrified rhizomes from the Miocene of Washington.- Amer. J. Bot., 69(1): 116-121.
Morgan, J. 1959. The morphology and anatomy of American species of the genus *Psaronius*.- Illinios Biol. Monogr. No. 27, University of Illinois Press, Urbana.
Morgan J. & T. Delevoryas 1952a. An anatomical study of *Stipitopteris*.- Amer. J. Bot., 39: 474-478.
Morgan J. & T. Delevoryas 1952b. *Stewartiopteris singularis*: a new psaroniaceous fern rachis.- Amer. J. Bot., 39: 479-484.
Nathorst, A.G. 1910. Beiträge zur Geologie der Bären-Insel, Spitzbergens und des König-Karl-Landes.- Bull. Geol. Instit. Upsala, 10: 261-415.
Nishida, H. 1986. A new *Tempskya* stem from Japan.- Trans. Proc. Paleontol. Soc. Jpn., N.S., 143: 435-446.
Nishida, H. 1989. Structure and affinities of the petrified plants from the Cretaceous of Japan and Saghalien, V. Tree fern stems from Hokkaido, *Paracyathocaulis orgurae* gen. et comb. nov. and *Cyathocaulis yezopteroides* sp. nov.- Bot. Mag. (Tokyo), 102: 255-282.
Nishida, H. & M. Nishida. 1983. On some petrified plants from the Cretaceous of Choshi, Chiba Prefecture VII.- Bot. Mag. (Tokyo), 96: 93-101.
Nishida, H. & K. Tanaka. 1982. Anatomical studies of *Cyathocaulis naktongensis* Ogura from central Hoshu, Japan.- Bull. Natl. Sci. Mus., Serc. C. (Geol. Paleont.), 8: 19-30.
Ogura, Y. 1933. On the structure of a fossil fern stem of *Cibotium*-type from the Upper Cretaceous of Iwate.- Bot. Mag. (Tokyo), 47: 748-754.
Ogura, Y. 1941a. On the structure of a fossil tree fern stem from Mindoro, Philippine Islands.- Jubil. Publ. Prof. Yabe; Contrib. Bot. Inst. Fac. Sci. Imp. Univ. Tokyo, 241: 911-917.
Ogura, Y. 1941b. Additional notes on the structure of fossil tree ferns.- Bot. Mag. (Tokyo), 55 (658): 453-461.
Penhallow, D. P. 1902. *Osmundites skidegatensis* n. sp.- Trans. Roy. Soc. Canada, II, 8: 3-18.
Petriella, B. 1969 *Menucoa cazaui* nov. gen. et sp., tronco petrificado de Cycadales. Provincia de Rio Negro, Argentina.- Ameghiniana, 6: 291-302.
Petriella, B. 1972. Estudio de maderas petrificadas del Terciario inferior del área central de Chubut (Cerro Bororo).- Rev. Mus. de La Plata (N.S.) (Palaeontologia No. 41), 6: 159-254.
Read, C.B. & R.W. Brown. 1937. American Cretaceous ferns of the genus *Tempskya*.- U.S. Geol. Surv. Prof. Pap., 186 F: 105-131.
Sahni, B. 1932. A petrified *Williamsonia* (*W. sewardiana* sp. nov.) from the Rajmahal Hills of India.- Palaeont. Indica, 20 (N.S.): 1-19.
Schelpe, E.A.C.L.E. 1955. *Osmundites natalensis* - A new fossil fern from the Cretaceous of Zululand.- Ann. Mag. Nat. Hist., 8 (12): 654-656.
Schelpe, E.A.C.L.E. 1956. *Osmundites atherstonei* - a new Cretaceous fern from Cape Province, South Africa.- Ann. Mag. Nat. Hist., 9: 330-332.
Schuster, J. 1911. *Osmundites* von Sierra Villa Rica in Paraguay.- Ber. Deuts. Bot. Ges., 29: 534-539.
Seward, A.C. 1907. Notes on fossil plants from South Africa.- Geol. Mag., 4 (5): 481-487.
Seward, A.C. 1910. Fossil Plants: A text book for students of botany and geology. Vol II. Cambridge University Press;

Reprinted 1963, Hafner Publishing Co., New York. 624 p.
Seward, A.C. 1917. Fossil Plants. Vol III. Pteridospermeae, Cycadofilices, Cordaitales, Cycadophyta.- Cambridge University Press; Reprinted 1963, Hafner Publishing Co. New York. 656 p.
Seward, A.C. 1924. On a new species of *Tempskya* from Montana, *Tempskya knowltonii*, sp. nov.- Ann. Bot., 38: 485-507.
Sharma, B.D. 1977. Indian williamsonias. An illustrated view.- Acta Palaeobot., 18: 19-29.
Smoot, E.L.; Taylor, T.N. & Delevoryas, T. 1985. Structurally preserved plants from Antarctica: I. *Antarcticyas* gen. n., a Triassic cycad stem from the Beardmore Glacier area.- Amer. J. Bot., 72: 1410-1423.
Snigirevskya, N.S. 1967. The remains of Calamites and Psaroniaceae ferns in coal-balls of the Donetzk Basin.- Palaeobotanica, 6: 7-25
Stidd, B.M. 1971. Morphology and anatomy of the frond of *Psaronius*.- Palaeontographica, B 134: 87-123
Stockey, R.A. & Smith, S.Y. 2000. A new species of *Millerocaulis* (Osmundaceae) from the Lower Cretaceous of California.- Int. J. Plant Sci., 161(1): 159-166.
Stopes, M.C. 1915. Catalogue of the Mesozoic plants in the British Museum (Natural History). The Cretaceous flora, Part 2, Lower Greensand (Aptian) plants of Britain.- British Museum (Nat. History), London, 360 p.
Tidwell, W.D. 1986. *Millerocaulis*, a new genus with species formerly in *Osmundacaulis* Miller (fossils: Osmundaceae).- Sida, 11: 401-405.
Tidwell, W.D. 1987. A new species of *Osmundacaulis* (*O. jonesii* sp. nov.) from Tasmania, Australia.- Rev. Palaeobot. Palynol., 52: 205-216.
Tidwell, W.D. 1991. *Lunea jonesii* gen. et sp. nov., a new member of Guaireaceae from the mid-Mesozoic of Tasmania, Australia.- Palaeontographica, B 223: 81-90.
Tidwell, W.D. 1994. *Ashicaulis*, a new genus for some species in *Millerocaulis* Erasmus ex Tidwell (Osmundaceae).- Sida, 16 (2): 253-261.
Tidwell, W.D. 1998. Common Fossil Plants of Western North America, 2nd Ed.- Smithsonian Institution Press, Washington, D.C., 299 p.
Tidwell, W.D. & Ash, S.R. 1990. On the Upper Jurassic stem *Hermanophyton* and it's species from Colorado and Utah, U.S.A.- Palaeontographica, B 218: 77-92.
Tidwell, W.D. & Ash, S.R. 1994. A review of selected Triassic to Early Cretaceous ferns.- J. Plant Res., 107: 417-442.
Tidwell, W.D. & Clifford, H.T. 1995. Three new species of *Millerocaulis* (Osmundaceae) from Queensland, Australia.- Australian System. Bot., 8: 667-685.
Tidwell, W.D. & N. Hebbert. 1992. Species of the Cretaceous tree fern *Tempskya* from Utah.- Intern. J. Plant Sci., 153(3): 513-528.
Tidwell, W.D. & Jones, R. 1987. *Osmundacaulis nerii*, a new osmundaceous species from Tasmania, Australia.- Palaeontographica, B 204: 181-191.
Tidwell, W.D., Kim, J.H. & Kimura, T. 1987. Mid-Mesozoic leaves from near Ida Bay, southern Tasmania, Australia.- Pap. Proc. Roy. Soc. Tasmania, 121: 159-170.
Tidwell, W.D. & Medlyn, D.A. 1991. Two new species of *Aurealcaulis* (Osmundaceae) from northwestern New Mexico.- Great Basin Nat., 51: 325-335.
Tidwell, W. D. & Nishida, H. 1993. A new fossilized tree fern stem, *Nishidacaulis burgii* gen. et sp. nov., from Nebraska-South Dakota, USA.- Rev. Palaeobot. Palynol., 78: 55-67.
Tidwell, W.D.; Nishida, H. & Webster, N. 1989. *Oguracaulis banksii* gen. et. sp. nov., a mid-Mesozoic tree-fern stem from Tasmania, Australia.- Pap. Proc. Roy. Soc. Tasmania, 123: 15-25.
Tidwell, W.D. & Parker, L.R. 1987. *Aurealcaulis crossii* gen. et sp. nov., an arborescent, osmundaceous trunk from the Fort Union Formation (Paleocene), Wyoming.- Amer. J. Bot., 74: 803-812.
Tidwell, W.D. & Pigg, K.B. 1993. New species of *Osmundacaulis* emend. from Tasmania, Australia.- Palaeontographica, B 230: 115-132.
Tidwell, W.D. & A.C. Rozefelds 1990. *Grammatocaulis donponii* gen. et sp. nov., a permineralized fern from the Jurassic of Queensland, Australia.- Rev. Palaeobot. Palyn., 66: 147-158.
Tidwell, W.D. & A.C. Rozefelds 1991. *Yulebacaulis normanii* gen. et sp. nov., a new fossil tree fern from south-eastern Queensland, Australia.- Australian System. Bot., 4: 421-432.
Tidwell, W.D. & Skog, J.E. 2002. Three new species of *Aurealcaulis (A. burgii* sp. nov., *A. dakotensis* sp. nov., and *A. nebraskensis* sp nov.) from South Dakota and Nebraska, USA.- Palaeontographica (in press).
Tidwell, W.D., L.S. Tidwell, W. Wright, & U. Dernbach 2001. *Tempskya* sp. from Neuquen, Argentina. The first *Tempskya* species reported from the Southern Hemisphere.- Amer. J. Bot., Abstr. 73.
Walton, J. 1923. On *Rhexoxylon* Bancroft, a Triassic genus of plants exhibiting a liane-type of vascular organization.- Phil. Trans. Roy. Soc., London, B 212: 79-109.
Watson, J. & C.A. Sincock 1992. Bennettitales of the English Wealden.- Mon. Palaeontograph. Soc., London., 558: 1-228.
Wieland, G.R. 1906. American fossil cycads.- Carnegie Inst. Washington Publ., 34: 1-196.
Zalessky, M.D. 1927. Flore Permienne des limites ouraliennes de l'Angaride.- Mem. Comm. Geól., n. sér., 176: 1-52.

Stockey: A Reinterpretation of the Cerro Cuadrado Fossil "Seedlings", Argentina

Burrows, G.E., Boag, T.S. & Stockey, R.A. 1992. A morphological investigation of the unusual crypogeal germination strategy of bunya pine (*Araucaria bidwillii*) – an Australian rain forest conifer.– Int. J. Plant Sci., 153: 503-512.
Burrows, G.E. & Stockey, R.A. 1994. The developmental anatomy of cryptogeal germination in bunya pine (*Araucaria bidwillii*).– Int. J. Plant Sci., 155: 519-537.
Calder, M.G. 1953. A coniferous petrified forest in Patagonia.– Bull. British Museum (Nat. Hist.) Geol., 2: 97-138.

Clarkson, J.R. & Clifford, H.T. 1987. Germination of *Jedda multicaulis* J.R. Clarkson (Thymelaeaceae): an example of cryptogeal germination in the Australian flora.– Aust. J. Bot., 35: 715-720.
Del Tredici, P. 1992. Natural regeneration of *Ginkgo biloba* from downward growing cotyledonary buds (basal chichi).– Amer. J. Bot., 79: 522-530.
Del Tredici, P. 1997. Lignotuber development in *Ginkgo biloba*. pp. 119-126- In: Hori, T.; Ridge, R.W.; Tulecke, W.; Del Tredici, P., Trémouillaux-Guiller & Tobe, H. (Eds.) *Ginkgo biloba* – a global treasure: from biology to medicine.– Springer-Verlag, Tokyo.
Del Tredici, P. 1998. Lignotubers in *Sequoia sempervirens*: development and ecological significance.– Madroño, 45: 255-260.
Gothan, W. 1950. Über die merkwürdigen feigenartigen Kieselknöllchen aus dem versteinerten Wald des Cerro Cuadrado in Patagonia.– Misc. Acad. Berol., 1: 149-154.
Menendez, C.A. 1960. Cono masculino de una conifera fósil de Bosque Petrificado de Santa Cruz.– Ameghiniana, 2: 11-17.
Nishida, M. 1981. A corm-like hypocotyle of araucarian seedling from the Upper Cretaceous of Hokkaido.– J. Japan. Bot., 56: 15-20.
Nishida, M. 1984. On the permineralized woods described in this research. pp. 3-20.- In: Nishida, M. (Ed.) Studies on affinities between Japanese Cretaceous plants and South American living plants. Chiba University, Chiba (in Japanisch).
Spegazinni, C. 1924. Coniferales fósiles Patagonicas.– Ann. Soc. cient., argent. Buenos Aires., 98: 125-139.
Stockey, R.A. 1975. Seeds and embryos of *Araucaria mirabilis*.– Amer. J. Bot., 62: 856-868.
Stockey, R.A. 1977. Reproductive biology of the Cerro Cuadrado (Jurassic) fossil conifers: *Pararaucaria patagonica*.– Amer. J. Bot., 64: 733-744.
Stockey, R.A. 1978. Reproductive biology of Cerro Cuadrado fossil conifers: ontogeny and reproductive strategies in *Araucaria mirabilis* (Spegazzini) Windhausen.– Palaeontographica, B 166: 1-15.
Stockey, R.A. 1982. The Araucariaceae: an evolutionary perspective.– Rev. Palaeobot. Palynol., 37: 133-154.
Stockey, R.A. 1994. Mesozoic Araucariaceae: morphology and systematic relationships.– J. Plant Res., 107: 493-502.
Stockey, R.A. & Taylor, T.N. 1978. On the structure and evolutionary relationships of Cerro Cuadrado fossil conifer seedlings.– Bot. J. Linn. Soc., 76: 161-176.
Stockey, R.A., Nishida, M. & Nishida, H. 1990. Structure and diversity of the woody conifer seedling-like structures from the Upper Cretaceous of Hokkaido, Japan.– Bot. Gaz., 151: 252-262.
Tidwell, W.D.& Medlyn, D.A. 1992. Short shoots from the Upper Jurassic Morrison Formation, Utah, Wyoming and Colorado, USA.– Rev. Palaeobot. Palynol., 71: 219-238.
Wehrfeld 1935. Patagonia, el gran acervo de fósiles de la Argentina.– Revista geográphica americana, Buenos Aires, 4: 117-130.
Wieland, G.W. 1935. The Cerro Cuadrado Petrified Forest.– Carnegie Inst. Wash. Publ., 449: 1-183.
Windhausen, A. 1924. Lineas generales de la constitución geológicas de la region situada al oeste del Golfo de San Jorge.– Bol. Acad. Cienc. Córdoba., 27: 167-320.

Selmeier: The Northern Alpine Molasse Basin - A Treasure Trove for Collectors of Petrified Woods

Jung, W. & Mayr, H. 1980. Neuere Befunde zur Biostratigraphie der Oberen Süsswassermolasse Süddeutschlands und ihre palökologische Deutung. - Mitteilungen der Bayerischen Staatssammlung für Paläontologie und historische Geologie, 20: 159-173, München.
Selmeier, A. 1985. Jungtertiäre Kieselhölzer (Bombacaceae, Tiliaceae, Meliaceae) aus dem Ortenburger Schotter von Rauscheröd (Niederbayern).- Münchner Geowissenschaftliche Abhandlungen, Reihe A, Geologie und Paläontologie, 6: 89-140, München.
Selmeier, A. 1991. Verkieselte *Castanea*-Hölzer aus dem Neuburger Wald bei Passau (Niederbayern).- Mitteilungen der Bayerischen Staatssammlung für Paläontologie und historische Geologie, 31: 149-165, München.
Selmeier, A. 1992. Verkieselte Eichenhölzer, *Quercoxylon* Kräusel 1939, aus jungtertiären Schichten des nordalpinen Molassebeckens.- Courier Forschungs-Institut Senckenberg, 147: 295-307, Frankfurt a. Main.
Selmeier, A. 1996. Tertiary *Platanus* woods from the northalpine Molasse basin (Austria, Germany).- Mitteilungen der Bayerischen Staatssammlung für Paläontologie und historische Geologie, 36: 157-183, München.
Selmeier, A. 1998. Aufsammlung von Kieselhölzern aus tertiären Schichten Süddeutschlands, der Schweiz und aus Österreich.- Mitteilungen der Bayerischen Staatssammlung für Paläontologie und historische Geologie, 38: 275-300, München.
Selmeier, A. 2000 a. Structural variation of Tertiary *Grewia* woods (Tiliaceae) from the East Bavarian Molasse, Germany.- Feddes Repertorium, 111: 465-480, Berlin.
Selmeier, A. 2000 b. A silicified *Pistacia* wood, Anacardiaceae, from the Upper Miocene of southern Franconian Alb, Germany.- Mitteilungen der Bayerischen Staatssammlung für Paläontologie und historische Geologie, 40: 201-212, München.
Selmeier, A. 2001. A silicified wood of Juglandaceae from the Lower Miocene sand and gravel pit at Rauscheröd, East Bavarian Molasse, Germany.- Mitt. Bayer. Staatssammlung Paläontologie und historische Geologie, 41: 95-109, München.
Selmeier, A. & Grosser, D. 1980. *Paradoxoxylon leuthardtii* Kräusel, eine Bennettitiaceae aus den pontischen Vogesenschottern des Delsberger Beckens (Schweiz).- Verhandl. der Naturforschenden Gesellschaft Basel, 89: 79-87, Basel.

Prof. Dr. Manfred Barthel

1934 born in Dresden, Germany
1952-1957 studies of geology, palaeontology, geography and botany at the University of Halle
since 1956 research work about Lower Permian plants from the Döhlen Basin (Saxony), later also from Thuringia
1960 doctorate. Thesis about epidermal structures of pteridosperms
1964-1968 Laboratory of Palaeobotany in the German Academy of Science, Berlin
Dr. habil. thesis about Tertiary ferns and cycads
1968-1992 Museum of Natural History at the Humboldt-University Berlin
1981-1992 Director of this museum
1992 retired, freelance.

Ulrich Dernbach

1939 born in Bad Salzschlirf, Hessen, Germany
1959 Abitur in Lauterbach
Work-study in pharmacy
Studied chemistry in Berlin
1965 joined the Merck Group (pharmaceuticals)
First fascination with fossils and emerging personal interest in palaeotology began in 1970.
Special interest and focus of his collection: petrified woods from all over the world, and, ever since his first trip to Argentina, araucaria cones from Patagonia
His books: Auraucaria (1992), Petrified Forests (1996).

Dr. Jean Galtier

1940 born in Montpellier, France
1957-1962 Licence Biology-Geology, University Montpellier
1963- Attaché recherche CNRS
1970 Dr ès Sciences, Montpellier University, dissertation on Lower Carboniferous plants from France
1975 Directeur recherche CNRS
Head of the Palaeobotanical Laboratory, Montpellier (1976-2000)
1996-2000 President International Organization of Palaeobotany (IOP)
since 2001 at Laboratoire Botanique & Bicinformatique, CIRAD, Montpellier.

Prof. Dr. Walter Jung

1931 born in Nuremberg, Germany
1951 Abitur from secondary school emphasizing classical languages
Studied natural science at Ludwig-Maximilians-University, Munich; specialized in botany, zoology, geography, chemistry and palaentology for university teachers
1959 doctorate, supervised by Prof. Mägde-frau, dissertation in palaentology Since 1962 active at the Institute for Palaentology and Historical Geology of the University of Munich
1967 full professorship in palaeobotany
1978 extraordinary professorship.

Prof. Dr. Hans Kerp

1954 born in Venlo, Netherlands
1972-1980 studies in geology at the University of Utrecht
His principal research subjects were stratigraphy and palaeobotany.
1986 Ph.D.
1980-1989 and from 1990-1991 appointed at the Laboratory of Palaeobotany and Palynology in Utrecht, first as a research assistant, since 1987 as a lecturer
1989-1990 appointed as research associate in the Department of Geology of the University of Pennsylvania, Philadelphia, U.S.A.
1991 Professor of palaeobotany at the Westfälische Wilhelms Universität in Münster, Germany
His research concentrates on Palaeozoic and Early Mesozoic floras, particularly on the palaeobiology and palaeoecology.

Robert Noll

1961 born in Grünstadt, Palatinate, Germany
1976 studied machine fitting at BASF Ludwigshafen
1979 employed by BASF
1994 active as mechanical engineer with BASF
Interested in geosciences (mineral and fossil collecting) for more than 25 years, early specilization in micromineralogy and photography, later involved with collecting and preparation of plant fossils
since 1997 hobby-researcher with regard to Permian palaeobotany.

Dr. Ronny Rößler

1965 born in Stollberg, Germany
1981-84 training as miner, SDAG Wismut, Gera
1987-92 studies in geology at the Freiberg University of Mining and Technology
1992-95 appointed as research assistant at the Freiberg University of Mining and Technology, working on stratigraphy, palaeontology and sedimentology projects for oil companies
1995 doctorate
since 1994 teaching in palaeobotany at Freiberg University of Mining and Technology
1995 Director of the Museum of Natural History in Chemnitz
2001 Editor of „The Petrified Forest of Chemnitz"
His research concentrates on the palaeobiology and taphonomy of petrified forests, palaeobotany of the Permian and palaeozoic arachnids.

Prof. Dr. Gar W. Rothwell

Professor of Environmental and Plant Biology, Ohio University palaeobotanist specializing in the pylogenetic radiations of major land plant groups and the evolution of plant reproduction, authored more than 150 scholarly papers and a text book in Palaeobotany,
currently the Ohio University Presidential Research Scholar in Biomedical and Life Sciences

1973 Ph.D. in Botany at the University of Alberta,
Isabel Cookson Award (1971) and Edgar T. Wherry Award (1993), past chairman and Secretary/Treasurer of the Palaeobotanical Section of the Botanical Society of America,
currently Secretary of the International Organization of Paleobotany (IOP)
His interests include origin and early evolution of land plants, of seed plants, of isoetalean lycophytes, of conifers, and of araceous monocots.

Prof. Dr. Alfred Selmeier

1923 born in Freising, Germany
1942 Abitur in Munich
studied biology, chemistry and geography at Ludwig-Maximilians-University in Munich and at the Technical University in Weihenstephan
1959 doctorate

Scientific collaborator at the Institute for Palaentology and Historical Geology of the University in Munich
Focus of research: identification of petrified plant remains,
Guest researcher at scientific institutions in France, China, the Philippines and the United States.

Prof. Dr. Ruth A. Stockey

1974 M.Sc. at Ohio University
1977 Ph.D. at The Ohio State University in the field of palaeobotany
1978 she moved to Canada where she is currently a Professor in the Department of Biological Sciences at the University of Alberta in Edmonton, Alberta Canada.
served as Chairman and Secretary/Treasurer of Palaeobotanical Section of the Botanical Society of America

1984 co-organizer of the 2nd IOP-Conference held in Edmonton,
teaching in palaeobotany and morphology of land plants and has an active program training both undergraduate and graduate students in the field of paleobotany
Research interests include the whole plant biology of fossil plants, phylogeny and evolution of conifers, flowering plants, fungi, and ferns, with a recent emphasis on structure and evolution of aquatic angiosperms and monocots.

Prof. Dr. William D. Tidwell

Born in Nampa, Idaho U.S.A.
Studied geology, botany, and palaeobotany at Brigham Young University, Provo, Utah and Michigan State University, East Lansing Michigan
since 1966 Professor of Botany (palaeobotany) at Brigham Young University
He has published numerous scientific articles, abstracts, and books including the renown "Common Fossil Plants of Western North America"

PD Dr. Volker Wilde

1956 born in Unna, Germany
1974 studies in Prehistory and later Geology/Palaeontology at the University of Münster
1981 Diploma-thesis at the Forschungsstelle für Paläobotanik in Münster
1983-1987 working on PhD-thesis (Eocene leaf flora of Messel) in Frankfurt am Main
1988-1991 research assistant at the Geological and Palaeontological Mueum and Institute Göttingen (Lower Cretaceous floras of NW Germany)
Returned to Frankfurt/M. in 1991 on a postdoc-position and took over his present position as head of the Palaeobotanical Section of the Forschungsinstitut Senckenberg in 1996.
Teaching palaeobotany at the University of Mainz since 1991.
Main research focus at present are Eocene floras, Devonian microfloras, and permineralized material from the Permian.

W. Walton Wright, MS

Born in 1939
B. Sc. at California State University Fullerton
M. Sc. at University of California Riverside
worked for the U.S. Forest Service, later for 16 years at the University of California Riverside, as senior ecologist for a planing firm, and as consultant doing management and restoration plans and evaluating impacts to the environment.
Interested in plant ecology, morphology and taxonomy.

Araucaria

The editor and publisher takes the reader along with him on a fascinating journey to the mysterious world of Patagonia.

In an exciting account of his travels, Ulrich Dernbach recalls his first encounters with the local gauchos and describes his expedition to the petrified forest – "Cerro Cuadrado Bosque Petrificado."

The photos that he took during his excursions vividly capture this sometimes barren but always fascinating landscape.

D'ORO Publishers

Schlehenweg 2
D-64646 Heppenheim
Phone: +49 / 6252 / 5661
Fax: +49 / 6252 / 68587

160 pages
Large format: 24 x 34 cm

ISBN 3-932181-00X

US$ 50.00

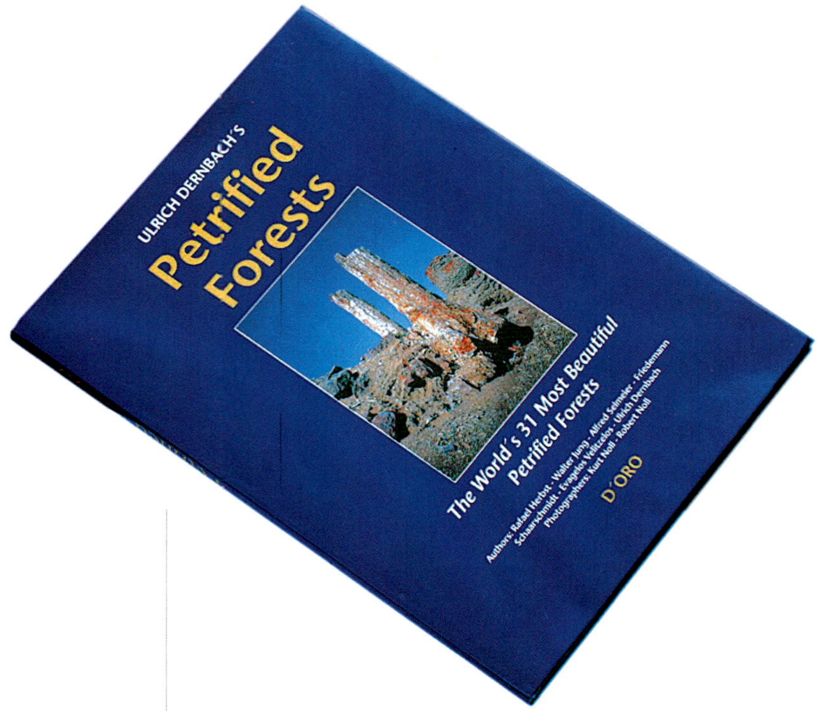

Petrified Forests

Ulrich Dernbach leads his readers to well-known sites where petrified woods are found, as well as to remote locations seldom before seen by human eyes.

Renowned palaeobotanists from around the world quickly agreed to his request to contribute papers detailing the findings of their many decades of research.

Including more than 300 color photos, *Petrified Forests* is an unprecedented synthesis of an opulently illustrated picture book, a captivating travel guide, and an instructive textbook.

D'ORO Publishers

Schlehenweg 2
D-64646 Heppenheim
Phone: +49 / 6252 / 5661
Fax: +49 / 6252 / 68587

190 pages
Large format: 24 x 34 cm

ISBN 3-932181-02-6

US$ 60.00